Business Process Improvement

The Breakthrough Strategy for Total Quality, Productivity, and Competitiveness

Dr. H. J. Harrington

The International Quality Advisor
Ernst & Young
San Jose, California

Sponsored by the American Society for Quality Control

McGraw-Hill, Inc.

New York St. Louis San Francisco Auckland Bogotá
Caracas Lisbon London Madrid Mexico Milan
Montreal New Delhi Paris San Juan São Paulo
Singapore Sydney Tokyo Toronto

I dedicate this book to my mother, Carrie. Throughout her life, she put two things before all others—hard work and me. As I look back, she received little reward for all she did. I wish I had done more.

Library of Congress Cataloging-in-Publication Data

Harrington, H. J. (H. James)
    Business process improvement : the breakthrough strategy for total
quality, productivity, and competitiveness / H. J. Harrington.
        p. cm.
    Includes bibliographical references and index.
    ISBN 0-07-026768-5
    1. Quality control.   2. Production management—Quality control.
I. Title
TS156.H338 1991
658.5′62—dc20   90-28544
                                                                      CIP

 This book is printed on recycled, acid-free paper containing a minimum of 50% recycled de-inked fiber.

          8 9 0   DOC/DOC   9 7 6 5 4 3

ISBN   0-07-026768-5

*The sponsoring editor for this book was James H. Bessent, Jr., the editing
supervisor was Olive H. Collen, and the production supervisor was Suzanne W.
Babeuf. It was set in Baskerville by McGraw-Hill's Professional Publishing
composition unit.*

Printed and bound by R. R. Donnelley & Sons Company.

Contents

Preface

We believe that there was one major quality breakthrough in the 1980s. It was not statistical process control, employee involvement, just-in-time, or total quality management. Nor was it quality function deployment, quality policy deployment, the improvement process, the Malcolm Baldrige Award, benchmarking, Taguchi methods, or poor-quality cost. It was the realization by management that the business and manufacturing processes, not the people, are the key to error-free performance.

The obvious is sometimes so difficult to see. As early as 1950, we realized that 80 percent of business problems could not be corrected by management. But we doggedly pursued statistical process control, in-process inspection, failure analysis, and employee involvement. We hoped that the new gimmicks, and even some old ones, would make our declining businesses winners again. In the name of quality, we threatened our suppliers, often forcing them to do things we were not doing ourselves. The result was the continuous erosion of our competitive advantage. The once-prestigious United States of America lost its title as the world's leading economic power and became instead the world's largest debtor nation. Our national debt boggles the mind of the average person, and continues to grow at a rate that worries the rest of the world. In short, the actions we took in the 1970s and 1980s fell far short of correcting the problems we had—and still have.

In the mid-1980s, a new concept emerged. It is something that we should have realized 30 years ago, when we were talking about management being responsible for 80 percent of the problems. Somehow we missed the true meaning of that realization. Think about it for a minute. Management doesn't run the drill press or operate a cash register. It doesn't write the software or assemble the computers. Maybe it should do a little of those jobs now and then to truly understand the process, but it doesn't. Dr. W. Edwards Deming, a well-known consultant, stated in 1986, "I should estimate that in my experience, most troubles and most possibilities for improvement add up to proportions something like 94 percent that belong to the system (process) and 6 percent to special causes."

Let me share an example of what I mean. Recently, in Helsinki, I stood in line at the ticket counter of a major U.S. airline for over 40 minutes. The ticket agents were working feverishly, but the procedure was slow and the line was long. The employees were not to blame for my wait. They were more unhappy about it than I was. In fact, four of them asked me to write a complaint letter to the company because, although they had discussed the problem with their manager, no one would do anything about it.

Management's job is to develop the business and manufacturing processes. The operators simply work within the boundaries of these processes. When you think about most of the problems we have, you will find that they start at the tip of the tongue, the point of the pencil, or the end of the phone line. Unfortunately, we have been working on the wrong end of the problem. What we have done, and very effectively I might add, is solve problems that never should have occurred in the first place.

We have fooled ourselves into believing that being the world's best problem solver would make us world-class, while what it really means is that we have had the most practice. The one place where American car manufacturers outshine the Japanese is in handling customer complaints. In the 1989 J. D. Power & Associates Customer Satisfaction Index, 9 of the 12 domestic makes were above average for fixing defects when they occurred. Chrysler was the benchmark car, resolving 61 percent of the owners' problems. When you have a lot of problems, you must get better at fixing them if you are going to keep your customers. That's the problem in the United States: We are spending all our time saying, "I'm sorry. I will fix it." when what we should be doing is developing processes that will make it unnecessary to ever say "I am sorry."

What we need to do now is prevent problems. "Prevention" is not preventing problems from recurring; it is preventing problems from occurring in the first place. We should avoid the thought pattern that says it is all right to make every error once. Many U.S. companies have developed a subconscious definition of quality as never making the same error twice. Although this is a step ahead of the attitude we had in the 1970s, when quality problems were solved only if the correction could be economically justified, we still have a long way to go.

As we enter the 1990s, our customers and our stockholders are not looking for good quality—they want perfection.

- *Quality* is doing the job right every time.
- *Perfection* is doing the right job right every time.

This is a new, more demanding environment, with a new type of customer. As Bert Staniar, chairman of Westinghouse Broadcasting Company, put it.

Today's customer is smarter, tougher, and less forgiving than ever before. Today, the customer comes prewired to be cynical, disloyal, and just plain ornery. He has been taught to demand quality, service, and greatness. He hears the words over and over again everywhere, and he's come to see it as his birthright.

This is a new breed of customer—one that evaluates his or her total interface with the organization, not just the product he or she receives. These customers are concerned about the way our salespeople react, how we package our products, how we advertise, how quickly we respond, the articles published about us, how we bill them, how our repair people communicate with them, the way our delivery people dress, and yes, even the political position our organization takes.

Customers' perception of our organization drives their buying habits. Combine this with an emphasis on total life-cycle cost, not just purchase price, and we have some very sophisticated customers to deal with.

The Forum Corporation reports that customers are five times more likely to switch to another supplier because of poor service than because of poor product quality or price issues. Customers will pay up to 30 percent more for an average product if they receive outstanding service from the organization. In fact, IBM's business strategy was built on the belief that people will pay more for outstanding quality and service. The 1970s and 1980s were the years in which American businesses were too busy swatting at problems to go out and spray the swamps. But customers in the 1990s will insist on a major change in our total business. Truly advanced organizations realize that we no longer can survive by pouring more resources into our business processes in an effort to overpower the competition. The antiquated business processes that we have lived with for years must be completely restructured, not just automated. Automating a bad process not only ensures that we can do a bad job every time but that we can do it faster and with less effort than before.

This new business strategy is changing management's thinking and action habits. Many of us are questioning some of the "sacred cows" (e.g., size of organization, span of control, centralized parts distribution centers, and plant locations) that we selected in the 1980s. For example, a company might be able to provide better customer service by having one large parts warehouse located at the hub of an overnight transportation carrier than by having a number of small warehouses scattered throughout the country. Do we really save money by having large manufacturing locations, or is it better to have smaller plants located close to the customer? We are beginning to realize that small business units of a maximum of 300 people provide the best return on investment.

Leading businesses are thinking differently about their processes. Processes are no longer viewed as just production processes. Today, management realizes that there are many more processes that use material, equipment, and people to provide many types of outputs and services. They are called *business processes*, and today they are even more important to competitiveness than production processes.

Throughout the 1980s, most companies focused their major efforts on correcting and improving their production processes. Only now is management realizing that it has been working on the wrong part of the business. The production process for an average product accounts for less than 10 percent of the product value, and the service industry that provides most of our jobs is 100 percent business processes. For years we have focused our efforts on measuring, controlling, certifying, and correcting our production processes. As a result, business processes became the major cost factor in our companies. The result was a new improvement strategy called *business process improvement*. Companies such as IBM, Corning, and Boeing have embraced this new approach—and realized some startling improvements including:

- Improved reliability of their business processes
- Improved response time
- Decreased cost
- Reduced inventories
- Improved manufacturability
- Increased market share
- Improved customer satisfaction
- Increased employee morale
- Increased profits
- Reduced bureaucracy

Everything we do today can be done better by concentrating on the process. Management has taken away our employees' ability to produce error-free output by saddling them with obsolete, cumbersome, bureaucracy-laden business processes. Let's give our employees a fair chance at success by starting business process improvement activities today.

Dr. H. J. Harrington

For CEOs Only

Why should you get caught up in these new quality and continuous improvement crazes? Why should you change your management style to be more participative? Why should you devote 30 percent of your personal time to be the quality role model? Why can't your quality assurance group take care of quality like it always has? The answer is very simple. The only reason you should ever consider starting an improvement process is to generate more profits and make your organization more competitive. The whole issue of quality should not be viewed by you from any other angle. Do it because it will make more profit for the organization than almost anything else you can do. It is the way to ensure the long-term success of your business. You need to look at the improvement thrust as a business investment that is going to either add to or detract from your long-term, net-favorable balance.

The question is, "Is this whole improvement process worth the investment?" The answer is a decided Yes, if it is managed right. Companies like IBM, Hewlett-Packard, Federal Express, 3M, and Xerox will testify to that. Companies like Globe Metallurgical brag that they are receiving a 40 to 1 return on every dollar invested in improvement. My experience indicates that a well-managed process will at least break even during the first 12 months, and from then on it's all "gravy."

The next question is, "How do you manage the improvement process correctly?" A good question, and one that is not asked often enough. Focus your improvement activities on improving every way you come in contact with your customers and/or potential customers, as well as the areas in which you invest large amounts of money. The mistake most organizations make is focusing the majority of their emphasis on the manufacturing process. In most businesses, manufacturing accounts for only 6 to 10 percent of the product cost. The sales force has much more of an impact on customer perception of your organization than most manufacturing activities. How many potential sales do they lose? There is a five times greater chance of losing a customer from poor service than from bad products. Just think about the losses your organization has suffered because market forecasts were bad, because products reach the marketplace late, or because the organization is having high personnel turnover rates.

The biggest opportunity you have to improve the bottom line comes from improving your business processes. In fact, your survival depends on improving them. In many companies, management can make more profits by cutting poor-quality cost in half than by doubling sales. This can be accomplished without hiring one new person, building one new building, or finding one new customer. There are dollar bills lying all around your organization. All you need to do is go out and pick them up and put them in the bank.

D. E. Petersen, CEO of Ford Motor Company, and H. A. Poling, its COO, wrote, "We are completely and thoroughly convinced that process improvement is vital, indeed critical, to our continued success." John Akers, chairman of IBM, points out, "Continuing to earn our reputation as a quality-intensive company is the best way I know of meeting our goals. To do so, we must also *improve our business processes.*"

Now comes the big question, "How do you improve your business processes?" A proven answer to that question can be found in the strategies presented in Chapters 1 through 10 of this book.

Dr. H. J. Harrington

Acknowledgments

I want to acknowledge the many contributions to this book made by the team at Harrington, Hurd & Rieker.

To Kelly Dorbandt, Kelly von Hoaglin, Lucy Lytle, and Cindy Reskovic, who converted and edited endless hours of dictation into the finished product.

To the following professionals, who helped develop the concepts:

- Dave Farrell
- Norm Howery
- Pravesh Mehra
- Francois Perrier
- Wayne Rieker
- Jean-Claude Savard

To Candy Rogers and Lee Dohaniuk, who kept things going and helped pull the book together.

To Dan Stowell for preparing the appendix and Bruce Sanders for writing the section on automation. To Armin Tietze and Rebecca Schroeder for their inputs to the BPI sophisticated tools. Their efforts significantly enhanced the end product.

But most of all, I want to recognize and acknowledge the contribution of my wife, Marguerite. She has stayed up late many nights and weekends proofreading, correcting grammar, and standardizing the format in this book. She was always there when I needed her.

It is important to state that the basic work on business processes started at IBM when I was still with the firm in the mid-1980s. This book expands upon this original work and reflects 4 years of applying these basic concepts to other organizations.

About the Author

H. James Harrington, MBA, Ph.D., is The International
Quality Advisor for the firm of Ernst & Young. He is also
president of the prestigious International Academy for
Quality, and the honorary advisor to the China Quality
Control Association. He is a much-sought-after international
lecturer and broadcasts his improvement methods regularly
on the Corporate Satellite Television Network.

Dr. Harrington's career in quality has been long and
distinguished, including 40 years with IBM, where he
became a senior engineer and project manager in Quality
Assurance. After leaving IBM, he became president of
Harrington, Hurd & Rieker Corporation. He served as
president and chairman of the board of the American
Society for Quality Control and as national vice president of
the International Management Council. In 1985, Dr.
Harrington was elected lifetime honorary president of the
Asia Pacific Quality Control Organization. He has been
elected honorary member of six quality associations outside
of North America and was installed in the Singapore
Productivity Hall of Fame.

He is the author of *The Improvement Process*
(McGraw-Hill), selected by *Library Journal* as one of the
best business books of 1986, and has written three other
business books published by the ASQC's Quality Press.

Why Focus on Business Processes?

INTRODUCTION

Americans have a problem, there is no doubt about it. Any way we measure our status, we are slipping backward. We will be the first generation in our history to have our children reach adulthood in a poorer economic environment than we enjoyed. The debt we are leaving them is staggering. Health care costs are out of control. Homeless people have become a major problem. Two million homeless roam our streets, and 14 million additional hidden homeless compound the problem. The American dream of owning our own homes has been priced out of reach unless there are two wage earners in the family. Our businesses are being bought by the Japanese at a worrisome rate.

What caused our living standards to decay? Why are Japan, Taiwan, South Korea, West Germany, and others getting fat eating off our plates? Many people say that it is because our production workers aren't working as hard as they should and are producing poor quality. I contend that the problem is much more complex and that the real issue rests within the white-collar, not the blue-collar, work force. The fact is, the productivity change rate for blue-collar workers is good. It is white-collar workers (e.g., engineers, bankers, lawyers, and managers) whose productivity is down. Between 1978 and 1985, the real output of blue-collar workers increased by 15 percent while their numbers decreased by 6 percent. The productivity gain was 21 percent. By contrast, the number of white-collar workers increased by 21 percent, but their real output increased by only 15 percent, for a productivity loss of 6 percent.

It is not the quality of their cars that is bringing Ford and General Motors to their knees. (Yes, the Japanese cars have better quality, with 1.19 problems per vehicle compared to 1.63 for American cars, according to J. D. Power & Associates' initial quality survey in 1989. But this difference is so slight that a customer would never notice.) It is the inability of the American automotive industry to provide new models featuring the latest technology in a reasonable amount of time.

Honda, Nissan, and Mercedes-Benz adjust the intake valve timing based on engine rpm, but this feature has yet to become available in U.S. vehicles. Electronically controlled automatic transmissions are the standard for Japan and Germany, but only Chrysler offers them in the United States. While Detroit relies on three-speed automatic transmission, Toyota and Mercedes-Benz offer a five-speed transmission. David Cole, the director of the Office for the Study of Automotive Transportation at the University of Michigan, states: "The Japanese set the benchmark in all details related to execution of the product. They can design, develop, and produce very complex products at very high quality levels quicker and less expensively than anyone."

Our national debt would be a fraction of what it is today, and our negative balance of trade would be drastically reduced, if white-collar productivity kept pace with blue-collar productivity.

Many companies believe that they cannot compete with the Asian market in the smokestack industries, so they are turning to service industries to find their niche. But what makes us think that we can possibly compete in the service industries if we can't compete in the basic product lines? The answer is, we cannot. We have nothing to offer that sets us apart in the service areas. The banking trends have proved that. During the 1970s, most of the top 10 banks in the world were American banks. Today, none are. Japanese banks have taken their place.

In November 1989, IBM opened a semiconductor research plant designed by a Japanese firm. Japanese companies are after our retail businesses also. They offered $1.3 billion to buy Bloomingdale's. Japan has brought together some of its best minds to develop a *fifth-generation computer technology* to leapfrog the U.S. software business. Direct foreign ownership of American business was up 21 percent in 1989.

Many of us are convinced that we cannot compete in the blue-collar areas and that our future lies in the service industries. If this is true, and I believe it is, we must drastically change the way we manage our *busi-*

ness processes—or we will run our country into bankruptcy at an even faster pace than we have been going.

WHERE HAS OUR REPUTATION GONE?

We have taken a fine worldwide reputation and destroyed it. The United States was once synonymous with high-quality products at reasonable prices. We lost this important customer advantage, and each day our reputation worsens because our competition is improving much more rapidly than we are. For example, the performance of American automobiles improved 8 percent in 1989. Japanese cars improved 17 percent, and European cars improved 21 percent (J. D. Power & Associates' 1989 survey). Consider the results of a survey conducted in the United States and Japan:

	American consumers (%)	Japanese consumers (%)
1. Which country has harder-working blue-collar workers?		
United States	10	3
Japan	51	85
Equal or no opinion	39	12
2. Which country has the most capable managers?		
United States	23	16
Japan	28	53
Equal or no opinion	49	31
3. Which country has the higher level of industrial technology?		
United States	34	26
Japan	26	41
Equal or no opinion	40	33
4. Which country makes the most advanced consumer products?		
United States	21	16
Japan	30	54
Equal or no opinion	49	30

Not only are we losing the respect of the rest of the world, we are losing respect within our own country. How can we hope to gain international market share when we cannot keep our customers here at home? It is much more difficult to rebuild a good reputation than to establish one. Customers happy with your output tend to stay with you.

They normally will forgive a mistake and give you a second chance. Only after repeated poor performance will they drop you in favor of the competition. Our lost reputation is caused by our long-term abuse and misuse of our previously happy customers. To keep from slipping further behind, we must make some very drastic changes in the way we understand and interface with our customers.

TODAY'S CUSTOMER

Do you really know your customers? Do you know what they need and expect? Are your customers excited about your products and services? If not, what will it take to excite them? Doing a good job won't do it. Doing a good job won't buy you customer loyalty. Show me one of your customers who believes that you are providing a good service that meets his or her needs, and I'll show you a customer who is a target for your competition. Good companies are on their way to bankruptcy, better companies are losing market share, and only the very best are going to grow in the future.

Your customer remembers your name under two conditions and two conditions only:

1. *When you provide extremely poor products or service.* "*FLY ME Airlines?* Oh yes, that's the airline that kept me waiting so long in line, and then lost my bags."

2. *When you provide surprisingly good products or service.* "Wow! That was a great meal and reasonable too. I need to get a matchbox so I won't forget the name. We should bring Mary and Joe here on Saturday."

To regain your lost reputation, you must provide *surprisingly good output* to your customers every time so that they become loyal customers. To rebuild a reputation and/or to increase your market share requires customers to start telling their friends and acquaintances that they are missing out on a good thing when they don't buy from you.

HOW CAN YOU PROVIDE SURPRISINGLY GOOD OUTPUT?

Customers today no longer take a microscopic view of your organization. There once was a time when you could build a good reputation by providing great products only. Today, however, customers view a potential supplier as a total entity. They expect every interface to be a pleasure. They expect the salesperson to be friendly and knowledgeable, the salesroom clean and pleasant, the bills readable and accurate, the package attractive and easy to open, the service people responsive and competent, the phones answered on the second ring and not to be put on hold. A 40-minute wait in a ticket line will be remembered long after the good service you receive on the plane. Good advertising frequently makes the difference between success and failure. In short, a surprisingly good customer experience is created only when every interface you have with your customer is orchestrated in a superior way. Nothing can be left to chance in the millions of moments of truth that occur when your customers come in contact with your organization.

To orchestrate these moments of truth, you must change your way of thinking, acting, and talking. You have to stop thinking about organizational structure and start focusing on the processes that control these customer interfaces. A completely different thought pattern occurs when you focus your emphasis on the process.

Organizational focus	Process focus
• Employees are the problem.	• The process is the problem.
• Employees.	• People.
• Doing my job.	• Help to get things done.
• Understanding my job.	• Knowing how my job fits into the total process.
• Measuring individuals.	• Measuring the process.
• Change the person.	• Change the process.
• Can always find a better employee.	• Can always improve the process.
• Motivate people.	• Remove barriers.
• Controlling employees.	• Developing people.
• Don't trust anyone.	• We are all in this together.
• Who made the error?	• What allowed the error to occur?
• Correct errors.	• Reducing variation.
• Bottom-line driven.	• Customer driven.

DON'T TRY TO MAXIMIZE PROFITS

American companies focus on profit to provide their stockholders with a better quarterly report and to maximize the executive team's annual bonus. What's wrong with that? It is the American way. If you maximize quarterly profits, upper management deserves to reap large rewards. The problem is that, in many cases, maximizing profits in the short run does not provide the best return on investment (ROI). Organizations that focus on building their reputations are going to provide the best long-term ROI. For example, during the past 20 years, companies in the auto industry that focused on reputation have outstripped companies that had a profit focus.

An organization that focuses on profit may have maximum profit in the near future, but an organization that focuses on reputation in the long haul will provide the best return to its investors. Unfortunately, American companies are focused on short-term profit, while Japanese companies work to improve their reputations. The Lexus LS400, a luxury sedan introduced by Toyota in the spring of 1990 to compete with world-class luxury cars, is an excellent example of a true luxury sedan. The Japanese are using a high-tech, four-valve, combustion-chamber engine and other advanced technologies to help build the car's reputation. Its real breakthrough is in price performance—a $50,000 to $60,000 car, selling for $35,000.

HOW DO YOU BRING ABOUT CHANGE?

Moving from an organization orientation to a process orientation is a difficult culture change. It requires a major change in the way the organization is managed. Change is not easy. Everyone is for change. Everyone thinks *he* should change, *she* should change, *they* should change, but *me* change? No way. Why do I need to change? I have proved that this is the right way to do things. Change is not a simple process. It requires a lot of thought, a well-developed plan, a sophisticated approach, and unfaltering leadership.

Here are 10 rules that should be used to guide your change process:

1. The organization must believe that change is important and valuable to its future.
2. There has to be a vision that paints a picture of the desired future state that everyone sees and understands.
3. Existing and potential barriers must be identified and removed.
4. The total organization must be behind the strategy to achieve the vision.
5. The leaders of the organization need to model the process and set an example.
6. Training should be provided for the required new skills.
7. Measurement systems should be established so that results can be quantified.
8. Continuous feedback should be provided to everyone.
9. Coaching must be provided to correct undesired behavior.
10. Recognition and rewards systems must be established to effectively reinforce desired behavior.

HARRINGTON'S CHANGE PROCESS CHART

Figure 1.1 illustrates the change process and how it reacts to time. Let me explain the various points on the chart:

P_1 = *Present condition.* When you start the change process, the existing condition is one in which the average performance is lower than you would desire, and there is a great deal of variation.

P_2 = *Preferred condition.* This is the condition that you want to achieve as a result of the change process. Normally, it provides better output to your customer at reduced costs and less variation.

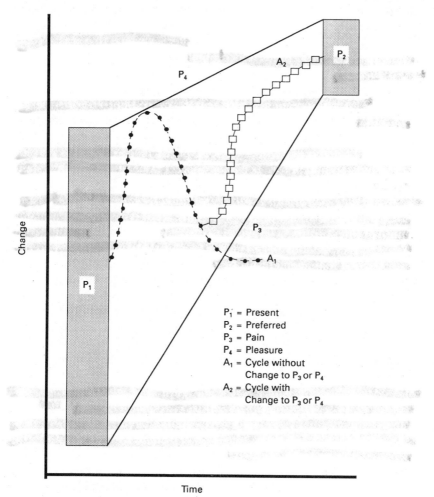

P_1' = Present
P_2 = Preferred
P_3 = Pain
P_4 = Pleasure
A_1 = Cycle without
    Change to P_3 or P_4
A_2 = Cycle with
    Change to P_3 or P_4

Time

Figure 1.1 Harrington's change process chart.

There is a time gap between the present (P_1) and preferred (P_2) condition points. Change does not occur overnight. It takes time to get it embedded into the fabric of the organization. The greater the magnitude of the change and the larger the organization, the more time it takes to bring about the desired change.

$P_3 = Pain.$ Even at the present condition, when performance falls below a certain level, the organization and/or the people in it begin to feel pain. People are scolded, transferred, demoted, or in extreme cases, fired. The organization loses prestige, money, customers, and sometimes even its stockholders. This can be thought of as the lower control limit. As the change process begins to take hold, this lower control limit needs to move upward. Minimum acceptable performance standards need to increase to show that the acceptable behavior level has changed. Performance that was acceptable in the past becomes unacceptable.

$P_4 = Pleasure.$ The P_4 line ties together the superior performance point and the present (P_1) and preferred (P_2) performance conditions. This is the point at which people are recognized for superior output. Here, too, as the change process continues, management must set increasingly higher standards for the outstanding performance. In some extreme instances, performance that was considered outstanding before becomes unacceptable under the new standards.

$A_1 = Average\ performance\ for\ a\ traditional\ education\ cycle.$ Note that during the initial phase of the education cycle, there is a major improvement in the average performance. However, it soon drops back to the previous performance level if there is no change in the supporting management process. There must be a close connection between when a student learns a new concept and when it is put to use in a real job application. In addition, the implementation of new concepts must be supported by a new set of performance requirements $(P_3$ and $P_4)$. If people do not apply the information presented to them in the first week after they attend a class, there is only a 20 percent chance they will ever use the techniques or methods taught.

$A_2 = Average\ performance\ when\ the\ pain\ and\ pleasure\ levels\ are\ changed\ in\ conjunction\ with\ an\ educational\ experience.$ When management helps the employees immediately apply their new knowledge and sets new performance standards to support the change, things really happen. The new performance standards $(P_3$ and $P_4)$ reinforce the lessons learned in class, thus helping the employees to change. Changing the pain and pleasure levels is an important reinforcement that brings about dramatic positive results.

WHAT IS A PROCESS?

There is no product and/or service without a process. Likewise, there is no process without a product or service. Before we go further, let me provide you with definitions of key words that will appear throughout this book:

System.   The controls that are applied to a process to ensure that it is operating efficiently and effectively.

Process.   Any activity or group of activities that takes an input, adds value to it, and provides an output to an internal or external customer. Processes use an organization's resources to provide definitive results.

Production process.   Any process that comes into physical contact with the hardware or software that will be delivered to an external customer, up to the point the product is packaged (e.g., manufacturing computers, food preparation for mass customer consumption, oil refinement, changing iron ore into steel). It does not include the shipping and distribution processes.

Business process.   All service processes and processes that support production processes (e.g., order process, engineering change process, payroll process, manufacturing process design). A business process consists of a group of logically related tasks that use the resources of the organization to provide defined results in support of the organization's objectives.

Organization.   Any group, company, corporation, division, department, plant, sales office, etc.

Function.   A group within a functional organization. Typical functions would be sales and marketing, accounting, development engineering, purchasing, and quality assurance.

Department.   A manager or supervisor and all the employees reporting to him or her.

Using these definitions, you can see that almost everything we do is a process and that business processes play an important role in the economic survival of our organizations.

The definitions are relatively simple, but most processes are not. Edward J. Kane, former director of quality for IBM Corporation, was very active in applying these concepts to the white-collar areas of IBM. He stated:

> Just taking a customer order, moving it through the plant, distributing these requirements out to the manufacturing floor—that activity alone has thirty sub-process steps to it. Accounts receivable has over twenty process steps. Information processing is a whole discipline in itself, with many chal-

lenging processes integrated into a single total activity. Obviously, we do manage some very complex processes separate from the manufacturing floor itself.

In all companies, there are literally hundreds of business processes going on every day. Over 80 percent of them are repetitive, things we do over and over again. I believe that these repetitive processes (white-, blue-, and gray-collar) can and should be controlled in much the same way as manufacturing processes are controlled. We manage many business processes that are as complex as the manufacturing process.

In the past, most of our attention was directed at process controls for the manufacturing area only. Today, the real payoff comes from applying proven manufacturing controls and feedback techniques to all activities in the business and treating the entire company as a complex operation containing many processes, only one of which is the process that produces the product sold to the customer. During the 1990s, there will be a major change in organizational philosophy, and management will begin to realize that it is management's job to lead a process revolution throughout the total organization.

A list of typical business processes defined by IBM follows. It will help you define your business processes.

Function	Process name
Development	Records management
	Acoustics control design
	Advanced communication development
	Cable component design
	Reliability management
	Cost target
	Design test
	Design and material review
	Document review
	High-level design specification
	Industrial design
	Interdivisional liaison
	Logic design and verification
	Component qualification
	Power system design
	Product management
	Product publication
	Release
	System-level product design
	System reliability and serviceability (RAS)
	System requirements
	Tool design
	User-system interface design
	Competitive analysis
	Design systems support

Function	Process name
Development (*cont.*)	Engineering operations
	Information development
	Interconnect planning
	Interconnect product development
	Physical design tools
	Systems design
	Engineering change management
	Product development
	Tool development
	Development process control
	Electronic development
	Phase 0 requirements
Distribution	Receiving
	Shipping
	Storage
	Field services and support
	Teleprocessing and control
	Parts expediting
	Power vehicles
	Salvage
	Transportation
	Production receipts
	Disbursement
	Inventory management
	Physical inventory management
Financial accounting	Ledger control
	Financial control
	Payroll
	Taxes
	Transfer pricing
	Accounts receivable
	Accrual accounting
	Revenue accounting
	Accounts payable
	Cash control
	Employee expense account
	Fixed asset control
	Labor distribution
	Cost accounting
	Financial application
	Fixed assets and appropriation
	Intercompany accounting and billing
	Inventory control
	Procurement support
	Financial control
Financial planning	Appropriation control
	Budget control
	Cost estimating
	Financial planning
	Transfer pricing
	Inventory control

Function	Process name
Financial planning (*cont.*)	Business planning
	Contract management
	Financial outlook
Information systems	Applications development methodology
	Systems management controls
	Service-level assessment
Production control	Consignment process
	Customer order services management
	Early mfg. involve. and product release
	EC implementation
	Field parts support
	Parts planning and ordering
	Planning and scheduling management
	Plant bus. vols. perf. management
	Site sensitive parts
	Systems WIP management
	Allocation
	Inventory projection
	New product planning
	WIP accuracy
	Base plan commit.
	Manufacturing process record
Purchasing	Alteration and cancellation
	Expediting
	Invoice and payment
	Supplier selection
	Cost
	Delivery
	Quality
	Supplier relations
	Contracts
	Lab procurement
	Nonproduction orders
	Production orders
	Supplier payment
	Process interplant transfer
Personnel	Benefits
	Compensation
	Employee relations
	Employment
	Equal opportunity
	Executive resources
	Management development
	Medical
	Personnel research
	Personnel services
	Placement
	Records
	Suggestions
	Management development and research

Function	Process name
Personnel (*cont.*)	Personnel programs
	Personnel assessment
	Resource management
Programming	Distributed systems products
	Programming center
	Software development
	Software engineering
	Software manufacturing products
Quality	New product qualification
	Supplier quality
Site services	Facilities change request
Miscellaneous	Cost of box manufacturing quality
	Service cost estimating
	Site planning

As we think about business processes, we need to recognize the difference between a process (how resources are used) and content (what resources are used). Consider a bowler, for example.

Process	Content
Selecting a bowling alley	Bowling alley
Finding someone to bowl with	Bowling partner
Putting on shoes	Bowling shoes
Picking out a ball	Bowling ball
Delivering the ball	Bowling pins, scoreboard, glove

PROCESSES VERSUS VERTICAL ORGANIZATIONS

To get economy of scale, most companies organize themselves into vertically functioning groups, with experts of similar background grouped together to provide a pool of knowledge and skills capable of completing any task in that discipline. This creates an effective, strong, confident organization that functions well as a team, eager to support its own mission. Unfortunately, however, most processes do not flow vertically; they flow horizontally (Figure 1.2).

A horizontal work flow combined with a vertical organization results in many voids and overlaps and encourages suboptimization, negatively impacting the efficiency and effectiveness of the process. Consider an order entry department that decided to stop comparing the item order number to the written description, even though it had been finding 3.3 percent errors. The department reasoned that it was the salesperson's

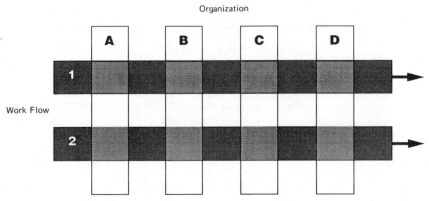

Figure 1.2   Horizontal work flow versus vertical organization.

responsibility to ensure that the order was filled out correctly, and it could use the extra 40 hours per week currently devoted to confirming the order to process the growing backlog of orders. This had a very positive effect on the order entry department's primary measurement. The number of hours required to process an average order dropped from 38 to 12 hours. Other benefits included:

1. Saving the time involved in making the comparison.
2. Saving the time and telephone cost required to contact the salesperson to determine whether the number or the narrative was correct.
3. By applying the time spent to check the orders to inputting orders into the computer, reducing overtime from 10 to 0 percent.

Upper management was so impressed with the improvement that it gave the whole department a "night on the town" at the company's expense.

Sounds great, doesn't it? Unfortunately, while the intention was good, the end results were disastrous. In this case, 2 percent of the customers began to receive the wrong product. The day before a major trade show, one customer, who had ordered 10,000 balloons with Ajax Tools printed on them, received 10,000 coats with Ajax Tools embroidered across them. The result was a spoiled promotional campaign for the customer—and it was very difficult for the supplier to sell 10,000 coats with Ajax Tools embroidered on them.

When you don't look at the total process, you have a group of individual small companies being measured on goals that are not in tune with the total needs of the business. This leads to suboptimization. Despite cases like these, a functional organization has many benefits, and a strategy is available to take maximum advantage of its effectiveness, as well as ensure that the processes provide maximum benefit to the company. That strategy is called *business process improvement* (BPI).

What we must do is stop thinking about the functional organization and start looking at the process we are trying to improve. I am always surprised when I sit down with a group of managers who are involved in a critical business process, such as accounts payable, and ask them, "Who owns the process?" Usually, I encounter eight different managers all sitting on their hands. In most cases, no one owns these critical parts of a business. Everyone is doing a good job, but no one makes sure that the activities interrelate. A critical part of BPI is to assign someone to own each critical business process.

Thomas J. Watson, Jr., former chairman of the board of IBM, explained the problem, saying,

> I believe the real difference between success and failure in a corporation can often be traced to how well the organization brings out the great energies and talents of its people. How does it help these people find common causes with each other? How does it keep them pointed in the right direction, despite the many rivalries and differences which may exist among them?

BPI ensures the effective and efficient use of resources—facilities, people, equipment, time, capital, and inventory.

MANAGING YOUR BUSINESS PROCESSES

You wouldn't consider hiring a group of production workers and telling them to go out and produce, without telling them specifically what you want them to do. Why is it, then, that we don't think we need to manage our business processes in the same controlled manner? In many organizations today, there are many individual groups all doing a good job. They are doing their own thing, very interested in meeting or beating their measurements, but not understanding or caring about how their activities affect others further down the process. They are interested only in what they are doing and how they are measured. This situation causes suboptimization to occur throughout the workplace. The three major objectives of BPI, then, are:

- Making processes effective—producing the desired results.
- Making processes efficient—minimizing the resources used.
- Making processes adaptable—being able to adapt to changing customer and business needs.

All well-defined and well-managed processes have some common characteristics:

- They have someone who is held accountable for how well the process performs (the process owner).

- They have well-defined boundaries (the process scope).
- They have well-defined internal interfaces and responsibilities.
- They have documented procedures, work tasks, and training requirements.
- They have measurement and feedback controls close to the point at which the activity is being performed.
- They have customer-related measurements and targets.
- They have known cycle times.
- They have formalized change procedures.
- They know how good they can be.

WHY FOCUS ON THE BUSINESS PROCESSES?

Expending much more effort to improve our business processes during the 1990s will be a major factor in being competitive in the twenty-first century. A focus on BPI helps the organization in a number of ways, by:

- Enabling the organization to focus on the customer
- Allowing the organization to predict and control change
- Enhancing the organization's ability to compete by improving the use of available resources
- Providing a means to effect major changes to very complex activities in a rapid manner
- Helping the organization effectively manage its interrelationships
- Providing a systematic view of organization activities
- Keeping the focus on the process
- Preventing errors from occurring
- Helping the organization understand how inputs become outputs
- Providing the organization with a measure of its poor-quality costs (waste)
- Providing a view of how errors occur and a method for correcting them
- Developing a complete measurement system for the business areas
- Providing an understanding of how good the organization can be, and defining how to get it there
- Providing a method to prepare the organization to meet its future challenges

Processes left unregulated will change, but that change will be for the convenience of the people in the process rather than for the best interest of the organization or the customer. Comfort and control, instead of effectiveness and prudent risk taking, become the rule.

THE HISTORY OF BUSINESS PROCESSES

Business processes were first developed out of a need to accomplish a specific business task. In most companies, these needs arose when the business was young and growing. They were developed quickly to meet an immediate need to serve a small internal population and a small customer base. After that meager beginning, they were neglected and ignored. They were not updated to keep pace with the business environment. No one took the time to review and refine them. As the organization grew, responsibility for these processes was divided among many departments, and additional checks and balances were put in place as small empires grew. Little pools of bureaucracy began to develop. One signature was replaced by two, three, or even four. Bureaucracy became the rule rather than the exception. Patches were put on top of patches. No one really understood what was going on, so no one could audit the business processes to ensure that they were operating correctly. Along the way, the focus on the external customer was lost. The organization became more inwardly focused, and people did not really understand the impact of their activities on the external customer.

Consequently, our business processes became ineffective, out of date, overly complicated, burdened with bureaucracy, labor intensive, time consuming, and irritating to management and employees alike. While most organizations accepted these processes as a necessary evil, they have turned out to be millstones around the organization's neck that increasingly hamper the ability of an organization to compete.

BUSINESS PROCESS FALLACIES

We need to ask ourselves why we allowed this to happen. There are a number of (false) beliefs that led management down this winding road to ineffectiveness. These fallacies include:

  • *Ineffective business processes do not cost the organization much money.* Wrong—Ineffective business processes are costing U.S. businesses billions of dollars every year. Between 40 and 70 percent of the white-collar effort adds no value. Eliminating white-collar errors and bureaucracy can cut overhead costs by as much as 50 percent, make your organization a leader in your field, and greatly improve your customers' perceptions of your organization.

  • *There is little to be gained by improving business processes.* Wrong—We already talked about the dollars and customers you save, but in addition, business process improvement can have the biggest single positive impact on the culture of your organization. No one likes to fight the many roadblocks we have placed in our business processes. People do not like to feel that their efforts are wasted. As these roadblocks are removed, morale will improve. Employees will stop being individuals and will become teams. Work will become fun, as little empires are crushed. The response time to internal and external customers will be cut in half. Studies at IBM in the mid-1980s showed that its productive sales time (time spent face to face with the customer) per salesperson had dropped to a small portion of what it had been just a few years earlier. Why? More and more administrative paperwork was the primary reason.

■ *The organization can work around business processes. Wrong*—The organization can work around your present business processes. You already have proven that. But do you want to work around them? Why have them if you are going to work around them? It is no wonder everyone is so hassled. Employees are so busy trying to find ways around the process that they don't have time to do their jobs. Tom spends huge amounts of his time making friends in purchasing so that he can get around the parts ordering process. His friend stops work that she is presently involved in to process Tom's request, without understanding how that affects the total process. Then purchasing employees wonder why they are spending all their time expediting hot jobs and setting inspection priorities in receiving inspection. They are in a whirlpool, and they see no way out. Sure, you can work around the business processes, but without a doubt, it is a bad business decision.

■ *Business processes cannot be controlled. Wrong*—Not only can they be controlled, they must be controlled. We need to stop missing schedules, losing customers, and fumbling business opportunities. Second-class business processes are only for second-class organizations. We need to control business processes much like we control manufacturing processes, to ensure high-quality results that will guarantee success.

■ *Business processes are unimportant compared to production processes. Wrong*—Customers are 5 times more apt to turn away from you because of poor business processes than poor products. Without a good interface between you and your customers, even the very best product will not attract them.

John R. Opel, chairman of the board of directors for IBM, stated,

> We need a business process that is worthy of respect and is respected. This means a process that can handle today's values and complexities accurately and efficiently—one that is positioned for the future and, therefore, can move ahead with the business, not struggle along behind as it has been doing.

WHAT HAS HAPPENED TO OUR BUSINESS PROCESSES?

Based on what we have discussed, some significant conclusions can be drawn:

- Business processes constitute a significant portion of your organizational costs.
- There is a significant opportunity to improve market share by improving your business processes.
- BPI enables you to make better business decisions and to implement them faster.
- BPI helps to improve and control your operations.
- BPI improves your production flow.
- Business processes have been ignored in the past.

Figure 1.3 depicts the way IBM views business processes. IBM found that, over the years, the business environment had become very complex, while business processes had changed very little, making them cumbersome and ineffective. The business processes had not kept up with the complexity of the business environment.

IS BPI FOR YOU?

When we talk about cutting your overhead costs by 50 percent, it sounds great. Everyone is for it. We all should be doing it right now. But don't jump too fast or start cutting your budgets too quickly. The change will not happen overnight. Most of the projects require many months of effort from people who are already very busy.

During the first 2 months, the process improvement team (PIT) members will work on the process about 160 hours each. Their commitment will then drop to about 10 hours per month for the next year, for a total of 280 worker-hours per PIT member. In addition, while this evaluation is going on, you will need to maintain your current business processes. Sometimes it is necessary to run parallel business processes (old process along with new process) to prove the effectiveness of the proposed change.

BPI will not happen by itself. It must be driven by top management. An executive improvement team (EIT) should be deeply involved in setting priorities for the business processes, appointing process owners, and reviewing progress. In addition, a BPI champion who understands BPI and can sell it to the organization is often assigned. It is an impor-

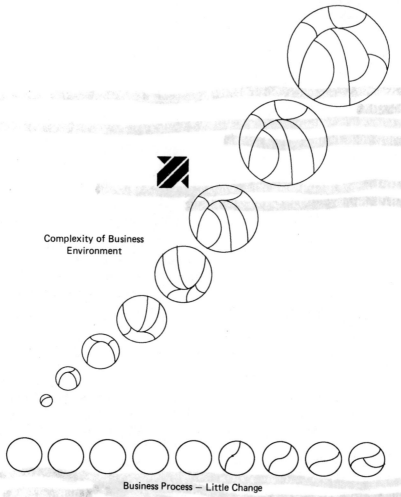

Complexity of Business Environment

Business Process — Little Change

Figure 1.3   IBM's view of the business process environment.

tant, logical change process, but it will not happen without some en-
lightened individuals to drive it.

WHAT IS BPI?

BPI is a systematic methodology developed to help an organization
make significant advances in the way its business processes operate. It
attacks the heart of the current white-collar problem in the United

States by focusing on eliminating waste and bureaucracy. It provides a system that will aid you in simplifying and streamlining your operations, while ensuring that both your internal and external customers receive surprisingly good output.

The main objective is to ensure that the organization has business processes that:

- Eliminate errors
- Minimize delays
- Maximize the use of assets
- Promote understanding
- Are easy to use
- Are customer friendly
- Are adaptable to customers' changing needs
- Provide the organization with a competitive advantage
- Reduce excess head count

THE FIVE PHASES OF BPI

The remainder of this book will discuss the five phases of BPI improvement (which are shown in Figure 1.4) and how to implement it within your organization. With the following methodology, you will be able to manage your business processes.

	Phase I. Organizing for improvement (Chapters 2 and 3)
Objective	To ensure success by building leadership, understanding, and commitment
Activities	1. Establish EIT 2. Appoint a BPI champion 3. Provide executive training 4. Develop an improvement model 5. Communicate goals to employees 6. Review business strategy and customer requirements 7. Select the critical processes 8. Appoint process owners 9. Select the PIT members
	Phase II. Understanding the process (Chapters 4 and 5)
Objective	To understand all the dimensions of the current business process
Activities	1. Define the process scope and mission 2. Define process boundaries

Phase II. Understanding the process (Chapters 4 and 5) (*Continued*)	
	3. Provide team training
	4. Develop a process overview
	5. Define customer and business measurements and expectations for the process
	6. Flow diagram the process
	7. Collect cost, time, and value data
	8. Perform process walkthroughs
	9. Resolve differences
	10. Update process documentation

Phase III. Streamlining (Chapter 6)	
Objective	To improve the efficiency, effectiveness, and adaptability of the business process
Activities	1. Provide team training
	2. Identify improvement opportunities:
	Errors and rework High cost
	Poor quality Long time delays
	Backlog
	3. Eliminate bureaucracy
	4. Eliminate no-value-added activities
	5. Simplify the process
	6. Reduce process time
	7. Errorproof the process
	8. Upgrade equipment
	9. Standardize
	10. Automate
	11. Document the process
	12. Select the employees
	13. Train the employees

Phase IV. Measurements and controls (Chapter 7)	
Objective	To implement a system to control the process for ongoing improvement
Activities	1. Develop in-process measurements and targets
	2. Establish a feedback system
	3. Audit the process periodically
	4. Establish a poor-quality cost system

Phase V. Continuous improvement (Chapters 8, 9, and 10)	
Objective	To implement a continuous improvement process
Activities	1. Qualify the process
	2. Perform periodic qualification reviews
	3. Define and eliminate process problems
	4. Evaluate the change impact on the business and on customers
	5. Benchmark the process
	6. Provide advanced team training

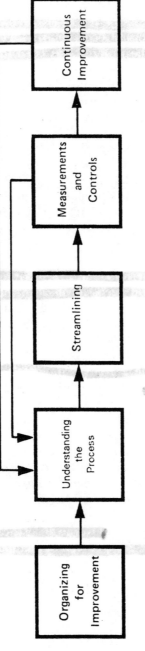

Figure 1.4 The five phases of BPI.

Organizing for Improvement

Understanding the Process

Streamlining

Measurements and Controls

Continuous Improvement

SUMMARY

We have been improving. Employee involvement, natural work teams, Quality Circles, and statistical process control (SPC) have led to continuous improvement, but the progress has been too slow. Many of the gains we made disappeared as soon as the focus was removed. In many industries we are still behind, and most businesses are feeling the heat of national and international competition. We can no longer be satisfied with improving at a gradual rate, or we will never be able to get ahead of the competition.

In the June 11, 1990, issue of *USA Today*, a major American auto corporation ran a full-page ad pointing out how, in 1980, its competition was 300 percent better than they were, and now, in 1990, the competition is only 25 percent better. Is it any wonder that this company has lost 30 percent of its market share? What do you think is going to happen to the remaining market share now that this large company has taken out ads notifying the potential customer that it produces poorer-quality cars than the competition?

What is needed in American industry today is a major change, a breakthrough that puts and keeps us ahead of our competition. This can be accomplished only by aggressively attacking both our manufacturing and business processes. BPI provides us with the means to quickly bring about a major improvement in the efficiency and effectiveness of all our processes and ensures that the gains that are made are maintained. A piecemeal strategy applied to improving processes will only create mediocrity. The first truth in process improvement is: If you bisect any process into its individual activities and then optimize the individual activities, the process as a whole will not operate as well as it could.

The business processes that most companies use have not kept pace with the business environment. As a result, organizations are wasting billions of dollars every year. Management must focus its attention on, and invest resources to revamp, the critical business processes that make companies efficient, effective, and adaptable to the needs of individuals, customers, and the organization. Overhead costs can be cut as much as 50 percent by applying these improvement methods to critical business processes.

With 75 percent of the U.S. work force in service industries, we must do something right now to make us more competitive. Our investment income balance is a sign of things to come. It peaked with a net surplus of $34.1 billion in 1981, dropped 93 percent to $2.3 billion in 1988, and went negative in 1989. In the November 6, 1989, issue of *U.S. News & World Report*, Citicorp chairman John Reed states, "The degree to which the United States improves its standard of living over the next 10 to 20 years depends very much specifically on the degree to which the service sector becomes more productive."

To improve your business processes, you must:

- Obtain management support
- Have long-term commitment
- Use a disciplined methodology
- Assign process owners
- Develop measurement and feedback systems
- Focus on the process

BPI is a prevention-oriented approach to managing the business. Solving problems may make things run better, but it does not bring about a long-term cultural change. To do that, you must change the processes that allow errors to occur in the first place.

2

Setting the Stage for Business Process Improvement

INTRODUCTION

In recent years, a new management truth has emerged. Companies pursuing quality as a competitive strategy have found that improved quality, increased productivity, reduced costs, and enhanced customer satisfaction go hand in hand. At the same time, they have found that the best way to ensure external customer satisfaction is to satisfy every internal customer at each step of the process. This condition is true for both manufacturing and nonmanufacturing operations.

In manufacturing settings, improvement can take many forms—new technology, better customer-supplier interfaces, the use of statistical controls, etc. Improvement in the white-collar areas and service industries takes the form of increasing the effectiveness and efficiency of the business processes that provide output to internal and external customers. Barriers that interrupt the flow of work must be removed, and the processes streamlined to reduce waste and lower costs. The best way to do this is through the concept called *business process improvement (BPI)*. Improving business processes is critical because overhead often represents 30 to 50 percent of the costs in a manufacturing organization and up to 80 percent of the costs in a service organization. This chapter provides you with an organizational model that will assist you in implementing BPI.

GETTING STARTED

Launching a BPI effort requires top management's support. By that, we do not mean that you have to have the CEO's participation, but you must, at a minimum, have the support of the head of the profit center that will be implementing the changes. (Often, profit centers that effectively implement BPI become the model for the total corporation when top management sees the results.) Of course, it is always best to start with the highest-level managers who will open their doors and their minds to you.

To get the BPI activities started, call a meeting of the business unit leader and the people who report to him or her to explain what BPI is and how it can benefit the organization. This meeting should cover:

- What poor-quality costs are in a typical business process
- An estimate of the organization's poor-quality cost in the business processes
- An overview of the BPI concept
- Sample business process problems currently facing the company
- An estimate of the cost of starting a BPI activity within the organization
- A request to approve the BPI concept
- A suggestion to form an executive improvement team (EIT)
- A request to appoint a BPI champion (czar)

This presentation will be a critical part of the total effort and in most cases should be conducted by an external consultant with experience in BPI to convince upper management to embark upon this journey. As you can see, a great deal of preparation is required for this meeting. The concept of poor-quality cost in the support and service areas needs to be understood, and some preliminary data collected. As a rule of thumb, poor-quality cost in the white-collar areas runs between 35 and 60 percent of their costs. In areas like development engineering, it runs as high as 80 percent.

FORM AN EIT

Most companies begin BPI by forming an EIT to oversee the improvement effort. If your company already has an EIT, the members should attend the kickoff meeting just described. If you don't have an EIT, now is the time to form one. The EIT should be chaired by the business unit head and consist of all the people who report to him or her.

The EIT's primary duty will be to manage BPI activities by:

- Communicating the need for BPI to the entire organization
- Releasing required supporting documentation (i.e., directives)
- Identifying problem processes in need of improvement
- Assigning business process owners
- Registering process improvement teams (PITs)
- Evaluating qualification level upgrade requests
- Following up to ensure that process improvement is an organizational priority
- Resolving conflicts that cannot be handled at lower levels
- Providing rewards and recognition to the members of successful PITs
- Measuring the success of the improvement effort

As we will see later, BPI efforts will involve several teams and subteams working on cross-functional processes and subprocesses. The entire effort has to be managed, coordinated, and controlled. Conflicts have to be resolved, and priorities set. All this is a top management activity which must be performed by the EIT. It cannot be delegated to anyone else in the organization. Visible and active support from the top could well be the difference between success and failure.

APPOINT A BPI CHAMPION

The next step is to appoint a BPI champion (often called a *czar*). This is necessary because BPI does not come easily, as change does not come easily. BPI is based on the belief that there is a better way of doing everything, and we have to find that way. This concept stimulates a feeling within an organization that things are *not* going right if they aren't changing. Instead of fearing change, people welcome it and think of it as a normal part of the business environment. In fact, after exposure to the BPI concept, people become uncomfortable when things aren't changing. They realize that a status quo condition is a state in which the organization and they, as individuals, are slipping backward.

Someone must be appointed to drive the BPI activities—a BPI improvement champion. The BPI champion's assignment will last for about 2 years, and he or she should report to the business unit leader. In large organizations, this will be a full-time assignment. In small organizations, it can be handled as a part-time assignment as long as the individual's workload is adjusted accordingly. Do not make it an additional responsibility for someone who already has a busy schedule. If your firm already has assigned someone to lead the improvement process, he or she also can serve as the champion for business process activities. The champion's job is to develop and customize the process improvement effort to your business and sell the approach throughout the organization.

Preventive activities always are difficult to start and even harder to keep going. The process will flounder if no one is steering the ship. The first time an emergency arises, work on the business process will stop and may never restart. The champion should provide the stimulus to keep the process moving no matter what and should serve as the EIT's eyes and ears.

The champion should prepare the business process directive and general job descriptions for the business process owners and PIT members. He or she will continuously review the individual teams' progress to determine when the process is ready to be qualified at the next higher level. The champion should have a detailed understanding of the BPI concept and all the tools that are used so that he or she can provide guidance to individual teams when they have trouble implementing BPI concepts.

Assigning a champion to lead the BPI activities demonstrates the head of the business unit's commitment to BPI. The champion should be selected carefully. He or she should be a person with stature, respected by the management team and the employees. Preferably, the champion should be at least in a functional management position and probably not from the quality assurance function. He or she should have high standards, believe the company can be better, embrace change, be a good salesperson, know how to lead teams, and want to take a leadership role in an activity that will have a long-term impact on the firm's business processes. This is an excellent assignment for a young, aggressive manager who has high potential to progress to top management or for a senior manager who wants to leave a permanent mark on the business before retiring.

One of the champion's first tasks will be to determine the scope of BPI activities as they relate to the organization. He or she then should develop, in conjunction with the EIT, procedures that define how BPI will be implemented within the organization. These procedures should be released under the business unit leader's signature.

In doing this, particular attention should be paid to the organization's structure and environment at different locations. Frequently, in major corporations, business processes acquire a global base and physically take on international perspectives that must be considered in the improvement model. For example, one of IBM's BPI activities addressed communication networks linking offices in 33 different countries. After

12 months of BPI, response time improved 300 percent, and system availability jumped from 86 percent to 95 percent.

The champion's position is not a permanent one; it should last about 2 years. By then, BPI activities will have become part of the management system and no longer will require a special person to maintain the momentum.

UNDERSTANDING THE PROCESS HIERARCHY

Almost everything we do or are involved in is a process. There are highly complex processes that involve thousands of people (for example, electing the president of the United States) and very simple processes that require only seconds of your time (for example, filling out your ballot). Because of these differences, we need to establish a process hierarchy (see Figure 2.1).

From the macroview, processes are the key activities required to manage and/or run an organization. New product definition is a good ex-

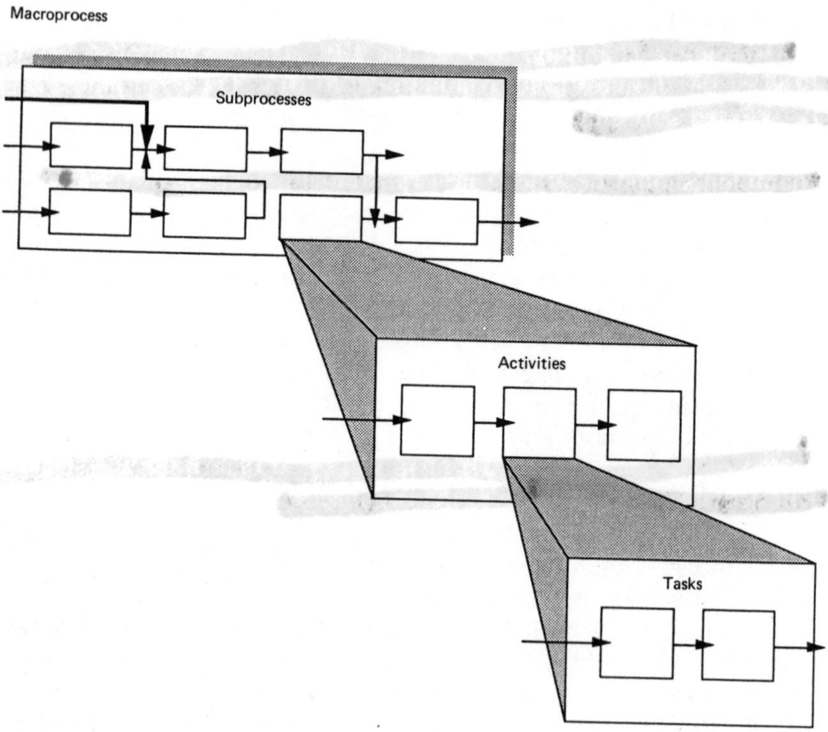

Figure 2.1   Process hierarchy.

ample of a macroprocess. Normally, a PIT is formed to work on improving a macroprocess.

A macroprocess can be subdivided into subprocesses that are logically related, sequential activities that contribute to the mission of the macroprocess. Selecting a presidential candidate is a good example of a subprocess of the macroprocess of electing a president of the United States.

Often, complex macroprocesses are divided into a number of subprocesses to minimize the time required to improve the macroprocess and/or to provide particular focus on a problem, a high-cost area, or a long-delay area.

Every macroprocess or subprocess is made up of a number of activities (for example, assessing the status of a meeting room to determine whether it is ready for a focus group meeting). Activities are things that go on within all processes. As the name implies, they are the actions required to produce a particular result. Activities make up the major part of flowcharts.

Each activity is made up of a number of tasks. For example, some of the tasks that are part of checking out the focus group conference room would be ascertaining that:

1. There are sufficient chairs for the invited guests.
2. Water and ice are in each of the pitchers.
3. A pad and pencil are placed on the table in front of each chair.

Normally, tasks are performed by an individual or by small teams. They make up the very smallest microview of the process.

EDUCATE THE EIT

Initially, the EIT should meet every 2 weeks to successfully launch the improvement effort. As this effort gains momentum, the number of meetings can be reduced to one a month to review the status of current projects and make new assignments. Such supervision typically will be required for the first year.

One of the EIT's key tasks is to become educated about BPI so that the team can lead the BPI concept and coach its managers and employees. The champion, frequently assisted by an experienced consultant, should design the educational workshop to:

- Familiarize the EIT with the specific purposes and activities of the BPI effort
- Build and reinforce the EIT's commitment to BPI

- Involve the EIT in analyzing and improving a critical business process so that the team members have a working knowledge of the concepts
- Motivate the EIT in launching and structuring the BPI effort

A 1-day workshop for the EIT should be scheduled early in the process. Most, if not all, EIT members will be top managers, who probably attended the first introductory BPI presentation. Consequently, the purpose of this workshop and discussion is not to unduly repeat but to ensure that the EIT is aware of what will happen within the organization, not just what could happen—as was the case in the first presentation. A typical agenda should include:

- An explanation of the workshop schedule and objectives
- An overview of the major BPI steps
- A review of the process of launching a BPI effort
- A review of the other major steps, providing only the essential information (e.g., goals, methods, challenges, and the support and contribution expected from the EIT at each step)
- A closer look at how to use the BPI tools

By now, the EIT should have a clear idea of what will happen in the organization. The second phase of the workshop should be an open group discussion to gauge people's reactions to BPI (i.e., their hopes, reservations, and feelings of resistance). A structured discussion should be built around questions such as:

- What makes us believe that we can succeed?
- What are the most important obstacles to BPI implementation in each manager's area?
- Where are the potential pitfalls?
- How do we expect our BPI champion to help us avoid these pitfalls?
- Are there activities with which we fundamentally disagree?

During this discussion, the following topics may be addressed:

- The critical business processes
- Outside case studies
- Educating the PITs
- Measuring success
- The vision statement
- The EIT mission statement
- Developing business process procedures

These questions may provoke reactions and resistance never before evident. The EIT chairperson and the BPI champion should be open and receptive to all comments. When problems are clearly and strongly stated up front, they are easier to manage afterward.

The group should end this discussion by having every person reassert his or her readiness to support the entire BPI strategy. Depending on the EIT members' interaction and the depth at which the BPI tools are presented, a second day may be required to get through the agenda.

LEADERSHIP BY EXAMPLE

Some organizations want the EIT to demonstrate its leadership by being the first PIT. This is an important step and shouldn't be rejected as "a job for people down the ladder." The advantages include:

- Enabling the EIT to experience firsthand the BPI activities
- Uncovering business processes that normally are not considered globally because they generally are managed through specialized functions (e.g., business planning, human resources management, and technology development)
- Learning to manage these processes from a process, not an organizational, perspective
- Providing demonstrated leadership
- Rapidly making major improvements in one of the organization's critical business processes

Presented correctly, this opportunity should motivate the EIT members and make them feel that by doing this assignment, not only will they learn something, but they also will be doing an essential part of their job—managing the organization's global processes. Obviously, the EIT chairperson is the first person to convince. Typically, he or she will be designated the owner of the process chosen by the group. Holding a meeting (prior to the workshop) between the improvement champion and the EIT chairperson should lay the groundwork for the commitment necessary here.

Another way to encourage EIT members to lead BPI activities is to name each of them the owner of one of the critical business processes. This approach provides better visibility to employees and positions the EIT members as leaders of the process.

ONGOING EIT ACTIVITY

The EIT should place the BPI effort on solid ground by tackling key management issues such as:

- Developing an operational model for BPI in the organization
- Reviewing implementation plans
- Drafting a BPI directive (or policy) for the organization
- Approving job descriptions for the process owners and PIT members
- Defining the business strategy and customer requirements
- Identifying the critical business processes existing in the organization
- Selecting processes to which BPI concepts will be applied
- Selecting the process owners
- Approving the criteria for registering the PITs
- Approving the qualification criteria
- Approving qualification level changes for the individual PITs

The EIT has several important decisions to make before starting BPI activities; hence, proper planning and documentation are essential. In our experience, generating documents (such as directives and plans) from scratch during the EIT meetings takes too much time. Instead, small subteams of two or three EIT members should be formed to prepare a draft version of documents that will be presented to the EIT for review and refinement. The BPI champion should coordinate these activities and develop and monitor a work plan that everyone agrees to. This is the best way to keep the process moving.

DEVELOP A BPI MODEL

Once management support and leadership for BPI exist, the EIT should develop a BPI model. This model is a detailed plan of the steps that will be undertaken as the organization goes through the BPI cycle. This will help management visualize the process, identify the sequence of events, and determine the resources necessary to implement the changes. This model is the design for your BPI activities, and the ultimate success or failure of your activities will depend on how well it reflects your organization's needs. As a result, considerable thinking and effort may be required to make the model reflect your corporate culture, resources, capabilities, and experiences. Chapter 1 of this book describes a typical model based on our experiences and should provide a good starting point for top management in developing its own model.

The basic five-phase approach of the typical model is:

1. Organizing for improvement
2. Understanding the process
3. Streamlining
4. Measurement and control
5. Continuous improvement

The activities involved in each of these phases were outlined in Chapter 1 and will be discussed in detail in the rest of the book. The model we described is a comprehensive approach for a company undertaking an organizationwide improvement process. An organization may decide to develop a scaled-down version for use in a pilot situation for a variety of reasons, ranging from inadequate resources, lack of faith in the methodology, or a need to better understand the potential return on investment. Although it is possible to scale down the process, the EIT should consider the fact that only limited improvements can be obtained from a limited scope.

IDENTIFY CRITICAL BUSINESS PROCESSES

The EIT is responsible for identifying the company's major existing business processes. In short, it should answer the questions, "What do we do as a business?" and "How do we do it?"

Two types of business processes typically coexist within an organization. One type is organized along functional lines, receiving its input from, and generating its output within, a single department. Such vertically aligned processes normally are very simple. They usually consist of a manager asking an employee to perform a specific task (e.g., to type a letter). Such simple processes normally are subprocesses of much more complex business processes, called *cross-functional business processes*.

Cross-functional processes flow horizontally across several functions or departments (see Figure 1.2 in Chapter 1). Usually, no single person ultimately is responsible for the entire process (e.g., design release or customer billing). All organizations have hundreds of business processes and thousands of subprocesses.

Top management should begin by listing only those business processes necessary to run the company. A typical organization's business processes might include:

- New product development
- Product design release
- Production planning
- Materials management
- Hiring
- Billing and collections
- After sales service
- Human resources training
- Customer needs analysis

We find that the most effective way to accomplish this task is to have each EIT member prepare a list of business processes he or she is involved in and submit this list to the BPI champion. The BPI champion will combine the lists into one overall list, eliminating the duplications. This combined list is then presented to the EIT members for their review and comments.

Overall, this is a good point at which to begin process analysis. However, for greater understanding, in some cases it will be necessary to hold another meeting of the EIT to break down complex processes into their subprocesses. The subprocesses of "after sales service," for example, might include:

- Repair and return
- Training of field repair personnel
- Field problem resolution
- Technical assistance
- Product servicing
- Warranty claims

Some companies supplement this top-down approach to identifying processes and subprocesses by also working from the bottom up, identifying specific activities, and then linking them to subprocesses. This ensures that no activities requiring significant resources are excluded from the analysis. This is a long, complex activity, and it should not be undertaken unless there is a specific need.

The EIT should decide what level of detail is appropriate: the process level, the subprocess level, or a combination of the two.

SELECT PROCESSES FOR IMPROVEMENT

Selecting the process to work on is a very critical step in the total BPI cycle. A lot of effort can be wasted, and the program can be dropped because of lack of interest and payback, if the wrong processes are selected. The processes that are selected should be ones where management and/or customers are not happy with the status quo. Normally, one or more of the following symptoms will be the reason for selecting a process for improvement.

- External customer problems and/or complaints.
- Internal customer problems and/or complaints.
- High-cost processes.
- Long cycle time processes.
- There is a better-known way (benchmarking, etc.).
- New technologies are available.
- Management direction based on an individual manager's interest in applying the methodology or to involve an area not involved otherwise.

Selecting these critical processes is one of the EIT's most important responsibilities. It never should be delegated. In selecting the process to work on, there are five things you should keep in mind:

- Customer impact: How much does the customer care?
- Changeability index: Can you fix it?
- Performance status: How broken is it?
- Business impact: How important is it to the business?
- Work impact: What resources are available?

Often, management gets carried away with its enthusiasm and hunger for improvement and overcommits its organization to the BPI activities. We suggest that you limit the initial phase to no more than 20 critical processes, although some companies have been very successful improving many more processes simultaneously. Martin Marietta started its process improvement activities in 1988, and by the middle of 1989, it had analyzed about 200 different processes and implemented changes to more than 125 of them. The following are a number of approaches that have been used to select critical processes.

Total Approach

Some companies' BPI efforts tend to occur on an organizationwide basis. Management opts to make improvements in all areas of the firm at once and launches a multitude of projects. This approach is feasible in a small organization. However, experience has shown that, unless the organization manages the improvement effort very effectively, such programs can be expensive and time consuming, the normal ongoing business is neglected, and the improvement effort tends to lack focus and coordination.

Management Selection Approach

The management selection approach is most often used to focus the improvement effort on the processes that are the most critical to future business success and on the problem areas identified by management. In this approach, the EIT develops a list of processes it considers critical to business performance. It then selects the 20 that are the most critical to the business. A second list, focusing on problem business processes, is prepared. Again, the EIT selects the 20 processes that are presenting the biggest problems to the organization and its external customers, based on its best judgment. Then these lists are combined. The resulting list will usually contain about 30 processes. The reduction from 40 to 30 occurs because some of the most critical processes appear on both lists. The prime candidates for BPI are the processes that appeared on both lists.

Weighted Selection Approach

Another way of accomplishing the same task is to have the EIT give each of the major business processes a rating (from 1 to 5) in the following four categories:

- Customer impact
- Changeability
- Performance
- Business impact

A rating of 1 indicates that the process is difficult to do anything with or that it has little impact. A rating of 5 indicates that the process is very easy to change or that it has a big impact. The ratings of the four categories for each business process are totaled, and these totals are used to set the priorities as shown in Figure 2.2.

The EIT then will decide which business processes should be addressed immediately and which can be scheduled for later analysis. Consideration should be given to balancing the work load within the organization and ensuring that all functions are participating in BPI. This approach concentrates attention on critical issues, sets priorities for resources, and ensures that the effort is manageable. While it is a relatively simple and useful way to select business processes, this approach has a number of drawbacks, including:

- "Pet projects" commonly are identified.
- Management perspectives may not be supported by hard facts.
- Top management may sway the decision.

Process Name	Changeability	Opportunities	Business Impact	Customer Impact	Total
Hiring Process	3	4	5	4	16
Accounts Payable	2	2	4	3	11
Engineering Change Release	5	3	5	3	16
Request for Quotations	4	4	4	3	15
Employee Appraisal Process	5	5	5	4	19
New Management Training	4	8	3	3	18

Figure 2.2   Setting business process priorities.

Informed Approach

World-class organizations continuously strive to provide superior products and services to their customers. As a result, they ensure that all improvement efforts focus on improving external customer satisfaction. Because a BPI effort is a major undertaking for all organizations which directly impacts the service-related customer interface, it should be linked to customer satisfaction.

Traditional approaches, as described earlier, can be thought of as *internally focused and defensive mechanisms* that try to correct and prevent problems from recurring. A more appropriate approach is *customer focused and offensive mechanisms* so that business processes are truly world-class. The informed approach we are going to describe in this section is an objective method for setting priorities for business processes, based on the importance of the process (as determined by external customer expectations) and the degree to which the process can be improved (as determined by its current quality). This approach embodies the following principles:

- Linking improvement efforts to customer expectations
- Focusing on prevention, as well as correction, activities
- Emphasizing the areas with the greatest potential for improvement
- Working on a manageable number of projects
- Using facts, not perceptions, for selection of projects
- Ensuring constancy of purpose

The informed approach differs from the other approaches because it is based largely on actual data collection from customers and internal operations, instead of on opinion. It is therefore more time consuming. The EIT should undertake this effort by:

1. *Understanding external customer requirements.* External customers have several requirements related to the product (e.g., features, durability) and the service that accompanies the product (e.g., delivery, order management, billing, after sales service). The service-related requirements are satisfied by the organization's business processes; therefore, it is essential to understand these requirements. This can be done by actually talking to customers and documenting their requirements or by using information that already exists in various parts of the organization (e.g., comment cards, customer service data). It is important to understand not only the requirements but also the importance of each of those requirements, since the aim of the organization should be to focus on the key requirements.

2. *Evaluating the importance of business processes.* External customer requirements are met by the organization through the execution of one or several business processes. This needs to be understood by

management and is best achieved by identifying the processes that directly and indirectly impact the external customer. Once the business processes that impact the external customer have been identified, the importance of these processes can be evaluated. Those processes that have major impact on important external customer requirements will be rated the highest and will be considered primary candidates for BPI.

3. *Evaluating the improvement opportunities.* No process is so perfect that it cannot be improved. Continuous improvement is the basic principle of BPI. However, the degree of improvement possible depends on the current state of the process. There are several key indicators of efficiency, effectiveness, and adaptability, such as cycle time, rework, backlog, and cost. The EIT should select a handful of meaningful indicators and then rate every process based on the information it can gather about the process. Processes with greater opportunities for improvement should be rated the highest, since they are primary candidates for BPI.

4. *Selecting the critical processes.* After gathering the data on the processes, the EIT is now ready to select the critical processes for improvement. It should be obvious that the critical processes are those that are high in importance, as well as high in improvement opportunities, and these are the ones that should be attacked first. Correspondingly, processes that are not very important from a customer viewpoint and appear to be functioning quite well should not be selected for initial improvement efforts. Figure 2.3 shows a matrix that can be constructed based on the data gathered. It is a helpful tool that provides a clear overview of all processes and helps make the final selection.

The same approach can be applied to improving the internal workings of the organization by replacing the external customer with the business unit. This is not as self-serving as it might seem. As you improve the internal workings of the organization, you reduce cost and provide a better quality of work life for the employees. As internal costs go down, the cost to your external customer can be reduced. As the quality of work life improves, the organization's output improves.

Remember that every process, every activity, every job within an organization exists for only one reason: to provide our customers and/or consumers with products and services that represent value to them. The rippling effect of improving any activity should have a positive impact on the external customer.

Assess Processes

Our assessment of the presented selection approaches is summarized on the next page.

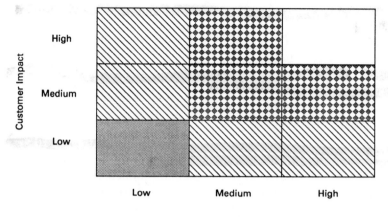

Improvement Opportunity

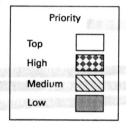

Figure 2.3 **Matrix for setting process priorities.**

Approach	Assessment
Total approach	Risky
Management selection approach	Practical
Weighted selection approach	Good
Informed approach	Best

As you complete your selection of the business processes that will have BPI applied to them, remember the "4 R's."

- *Resources.* There is a limited amount of resources available, and the present processes must continue to operate while we are improving them. Often, this means that a new process will be operating in parallel with the old process while the new process is being verified. Don't overextend yourself.

- *Returns.* Look closely at the potential payback to the business. Will the process reduce costs? Will it make you more competitive? Will it give you a marketing advantage?

- *Risks.* Normally, the greater the change required, the greater the risk of failure. Major changes always are accompanied by resistance to change. Breakthrough activities have the biggest payback but also have the biggest chance of failure.
- *Rewards.* What are the rewards for the employees and PIT members working to improve the process? How much will their quality of work life be improved? Will the assignment be challenging and provide them with growth opportunities?

Based on the results of this analysis, the EIT typically may select 10 to 25 critical processes to which BPI will be applied. The number will vary based on the size and complexity of the organization. At IBM San Jose, 86 of 250 business processes were selected initially as targets for BPI, an unusually high number to work on at one time and requiring a staff of five. We recommend that even a large organization address no more than 25 business processes during the initial phase.

PRELIMINARY OBJECTIVES

Once the EIT has selected the processes to which BPI will be applied, the EIT should develop a set of preliminary objectives that will be used to provide vision and direction to the PITs. These preliminary objectives will ensure that an initial common understanding exists between the PIT and the EIT. Depending on the amount of knowledge and the data that the EIT has about the selected processes, these objectives will be more or less quantified. In some cases, the objectives may only set the direction for the PIT (for example, reduced cycle time). In other cases, improvement objectives may be provided (for example, reduce cycle time by 20 percent). The preliminary objectives should address effectiveness, efficiency, adaptability, and cycle time. In all cases, the objectives should be focused on meeting or exceeding customer expectations and can drive incremental or breakthrough improvement, depending on the degree of improvement desired.

It is very important not to blindly accept the preliminary objectives as the goals for the PIT. Normally, these objectives are set without detailed data or understanding of the present process. Letting the PIT set its own goals provides needed ownership. These goals often are more aggressive than the EIT-set objectives.

OPERATING ASSUMPTIONS

Besides developing objectives, the EIT sometimes develops operating assumptions to help guide the PIT. These operating assumptions define the constraints (for example, employee resources), specific opportunities (for example, evaluate replacing the present phone system), or expectations (for example, all changes will be implemented within the next 12 months). Often, the EIT will develop a set of general assumptions that will apply to all PITs, and additional ones to provide direction to individual PITs. It is recommended that the EIT-generated assumptions be kept to a minimum since they may restrict the creativity of the PIT. The process owner and the PIT will further develop the operating assumption list as the process develops.

MANAGING BPI COMMUNICATIONS

BPI significantly alters the way we approach our organizations and the way we do business. This drastic change requires clear and direct top management communication to all employees to explain the new focus on business processes. People in all areas of the organization need to understand their roles in refining the processes and understand that error-free performance can be accomplished only by focusing everyone's efforts on process improvement.

One of the best ways to accomplish this is to issue a BPI message clearly stating management's commitment to BPI and outlining the role every employee will play in this important undertaking. Consider, for example, a message issued by IBM president John R. Opel in 1984:

> We need a business process that is worthy of respect and is respected. This means a process that can handle today's volumes and complexities accurately and efficiently...one that is positioned for the future and, therefore, can move forward with the business, not struggle along behind as it has been doing.

Another message, sent a year later by the new president of IBM, John Akers, authorized implementation of the business process concept, designated it a priority activity for the corporation, and expressed his personal commitment to process management. In part, it read:

> When you follow the multiplying effects of poor quality in the business process, you begin to see how much it costs in time...and, most important, in customer goodwill....We have to do things right the first time. That applies to cutting an invoice, processing a purchase order, or writing a memorandum, just as much as it does to manufacturing a product.

On April 21, 1988, Ford Motor's chief executive officer, D. E. Petersen, and chief operating officer, H. A. Poling, jointly released a letter on process improvement. Key statements in this letter were:

> We consider its [process improvement] successful implementation critical to achieving and maintaining our goals of leadership in quality, cost and on-time programs.
>
> It is with enthusiasm and our own personal commitment that we enlist your support in this mission of process improvement. We are completely and thoroughly convinced that process improvement is vital, indeed critical, to our continued success. It will be your dedication and commitment to your operation or activity that will generate success. Start today to evaluate use of your own time to ensure that this concept is assigned high priority.

The BPI champion should prepare a BPI directive and review it with the EIT. After approval, the business unit or division leaders should all sign this directive, and it should be distributed to the entire work force. The contents of this directive typically include:

- The need for improvement
- The concept of business processes
- The approach the company is taking
- Individual and group responsibilities
- Process qualification criteria

These key points should be disseminated to all employees through conventional communication systems, such as newsletters, weekly department meetings, and training seminars. First-level supervisors or managers should review the directive with all employees. To enhance the review process, the BPI champion should prepare a set of frequently asked questions and answers about BPI to distribute with the directive. On May 14, 1985, IBM's key BPI document was released, signed by W. W. Eggleston, then corporate vice president of quality. In part, it stated,

> Each IBM operating unit will apply quality management to its key functional and cross-functional processes. Line management will define and own these processes. They will have responsibility for, and authority over, the process results. An executive is named as the single process owner and must operate at a level high enough in the organization to:
>
> - Identify the impact of new business direction on the process.
> - Influence change in policy/procedures affecting the process.
> - Commit a plan and implement change for process improvement.
> - Monitor business process efficiency and effectiveness.

There is no one most effective approach to communications, so use a number of different approaches to be sure that everyone gets the word. The point is that top management must set the direction, provide the necessary resources, and communicate the message to the people who will be assigned to improve the process.

SELECT PROCESS OWNERS

The process owner is the individual appointed by management to be responsible for ensuring that the total process is both effective and efficient. This is a key concept that must be understood to make the BPI strategy work. Conventional functional management has worked well for a number of years and is probably the best type of organization, but it has its shortcomings. Functional competition, although healthy in some cases, can be self-defeating since it puts all the functions in competition for limited resources. Frequently, the organization that puts on the best show gets the most resources, but it may not have the biggest need. In other cases, resources are allocated to part of a critical process by one function, but interfacing functions have different priorities and, as a result, only minor improvements are made.

What needs to be done is to stop looking at the business as many large functions and start looking at it as many business processes. This allows the organization to select the process it wants to improve, obtaining the maximum return on its investment. It is very evident that the process is the important element and that the process owner plays a critical role in making the total process mesh. The process owner concept provides a means by which functional objectives can be met without losing sight of the larger business objective.

The process owner must be able to anticipate business changes and their impact on the process. The owner must be at a high enough level to understand what direction new business will be taking and how it will impact the process.

As Edward J. Kane, former director of quality for IBM Corporation, put it,

> In IBM's case, to accomplish these things, the process owner becomes the focal point of a new and permanent integrating structure. It introduces a type of matrix management (but is driven and controlled by line management) for the functions that operate within the process. The owner may require a "do" function (project office) which has representation from the functions/units in the process. These individuals would be assigned by owners of the critical sub-processes. They would provide functional expertise to the process owner and be the implementers of change and simplification within their sub-process. The owner is the voice of process capability to the business (up) and for process implementation of change (down) within the existing line management structure.

Although the EIT may ask the BPI champion to propose candidates, selecting the process owners is an EIT responsibility.

Perhaps we should briefly explain the choice of the word *owner*. Some companies use the term *process manager* for the role. Although the term is basically correct, *manager* connotes responsibility for hiring, supervising, and signing paychecks. A process owner has no such supervisory responsibility. Another common term is *process leader.* While not inappropriate, the word *leader* has so many connotations that we prefer to use it with great care. Ford uses the term *process sponsors.* Ultimately, the BPI champion and the EIT should select a term appropriate to the organization's culture and accepted use of job titles.

Criteria for Selecting Process Owners

Ownership. Business processes seldom improve, because there isn't anyone who really feels that he or she owns them. Therefore, the first criterion must be ownership. One way to decide who feels (or should feel) the most ownership of a particular process is to answer the following questions. Who is the person with the most:

- Resources (people, systems)?
- Work (time)?
- Pain (critiques, complaints, firefighting)?
- Actual (or potential) credit?
- To gain when everything works well?
- Ability to effect change?

The answer to these questions should give a fairly good idea of who is the most concerned, and the most involved, with the process. Under some situations, the end customer of the process may be the best owner because he or she has the most to gain from its improvement.

Power to Act on the Process. A second issue to consider in process owner selection is that the critical processes identified by the EIT may come from varied organizational levels (corporate, divisional, regional, etc.). Therefore, the EIT must ensure that the owner has sufficient power to act on the selected process. Because many major business processes are interfunctional or even international, most of them do not have an organizational structure leader. Consequently, the EIT must give the process owner authority and responsibility for the total process.

The business process owner should be an individual who operates at a level high enough in the organization to:

- Identify the impact of new business directions on the process
- Influence changes in policies and procedures affecting the process
- Commit to a plan and implement changes
- Monitor the effectiveness and efficiency of the process

Leadership Ability. A third criterion for process owner selection concerns the person's ability to lead a team. He or she should be:

- Perceived as highly credible
- Able to keep a group on schedule
- Able to lead and direct a group
- Able to support and encourage PIT members in their improvement efforts
- A skilled negotiator
- Willing to embrace change
- Able to deal with higher-level management
- Able to knock down roadblocks
- Able to see the big picture
- Unafraid to take risks
- Able to live up to commitments
- Able to handle poor performers

Process Knowledge. The final criterion is that the process owner should have a good understanding of the process. If the process owner understands the total process, it is much easier for him or her to do the job, and there will be very few restarts. This is a desirable characteristic but is not mandatory, because as soon as the process is flowcharted, every member of the PIT will have a good understanding of how the total process works. As a result, a customer of the process who has little understanding of the internal workings of the process can be selected as the process owner.

Analysis of Criteria

The first two criteria, ownership and power to act, are of the utmost importance. If one of these cannot be met by a specific candidate, the success of future improvement activities is endangered. The right person for the job must be high enough in the organizational structure to have power, have enough time to become involved, and provide the necessary leadership.

The third criterion, leadership ability, should be used with good judgment. Indeed, a talented, well-respected, and assertive individual often can lead very well—sometimes better than a participative, but unassertive, individual.

The process in question should suggest the appropriate position in the organization from which the owner should come. If the person in that "natural" position does not fit the process owner role, because of personal attitudes or past reputation, train him or her in the necessary skills. If a process is crucial to your business, you will find a way to ensure effective ownership and leadership.

Authority can be given, asserted, or earned. Selecting a process owner who has the right combination of earned peer and top management respect will do more than any amount of documentation.

In most cases, being a process owner is not a full-time job. Although, initially, it may require additional work, as the process improves, the regular work load should decrease because the owner no longer must react to as many unscheduled interruptions. If the owner's initial work load is too great, this provides an excellent opportunity to delegate some of the less important work and increase employee involvement.

CONSIDER A PROCESS COORDINATOR

Because of the many demands on the process owner's time and his or her level in the organization, it may be inappropriate for the process owner to do all the follow-up work related to directing the PIT. In these cases, the process owner should appoint a process coordinator to work with him or her on the assignment. In a very large process, this could be a full-time job during the initial stages of the process.

CONSIDER A PROCESS IMPROVEMENT FACILITATOR

Depending on the maturity of your team structure and the PIT's knowledge of the BPI tools and methodology, it may be advisable to include a process improvement facilitator as part of the PIT. The facilitator should have an in-depth understanding of the BPI tools and experience in how to implement them. The facilitator does not need to have previous experience in the process under investigation. The facilitator's job is to guide the PIT through the initial planning and implementation stages. Often, an external consultant is used because he or she provides a fresh, unbiased view of the process, with no preconceived opinions and no internal organizational loyalties. Frequently, this neutral person is the catalyst who makes breakthrough possible.

PROVIDE JOB DESCRIPTIONS FOR PROCESS OWNER AND PIT MEMBERS

Launching a BPI effort is a major event for any company. It is a decision that will impact habits, attitudes, systems, and even technologies.

BPI is more than *adapting* old patterns to a new fashion; it fundamentally *changes* the way a firm does business.

Organizational change occurs only if the authority for installing that change exists. Consequently, one of the BPI champion's and the EIT's first duties will be to formally authorize some of the employees to improve the business processes. This will be accomplished most readily by having job descriptions prepared for the process owner and the PIT members. While these are not full-time jobs, they play such central roles in the improvement process that they require formal recognition by the EIT and the BPI champion. Of course, their performance in these roles should be evaluated during their annual performance evaluations.

Process Owner

General Scope of the Job. The job description for a process owner should indicate that he or she ultimately is responsible for improving a specific process. Indeed, management now should expect the owner to take all the actions necessary to ensure that the effectiveness and efficiency of the entire process are improved.

Being a process owner is comparable to being a program manager. A program manager usually has very specific goals (i.e., to deliver a new product by a certain date, in conformance to customer requirements). The business process owner's goal is to improve the assigned process to the point at which it reaches best-of-breed status and to keep it at that level.

John Akers, chairman of the board of directors of IBM, said,

> IBM has always been known for the high quality of its management. You've earned a reputation for being professional, energetic and sensitive. There's another trait IBM managers show that really sets you apart. That's the way you approach your work: not just as highly skilled professionals, but as "owners" of a business who know they must take responsibility for every aspect of their job.
>
> In no area is this attitude more important than in the way we *run our business*. I'm talking about things like procedures, paperwork, accounting, and controls ... all the activities we lump together under the term, "business process." In a large and complex organization like ours, the need is especially great for managers to "own" that process.

Major Responsibilities. The process owner's job description should include the following responsibilities:

- Establish measurements and set targets to improve process effectiveness, efficiency, and adaptability

- Ensure that the overall goals of the process are met and that the improvements made within the process do not negatively affect other processes or other parts of the organization (*suboptimization*)
- Define the preliminary boundaries and scope of the process
- Form a PIT by:
  Meeting with involved department heads to gain their commitment
  Obtaining the names of potential members
  Selecting team members
- Ensure that the PIT members are trained in BPI techniques and that they use basic improvement principles
- Launch the PIT's activities by helping to:
  Define process boundaries
  Establish the team's mission
  Register the team
- Organize the PIT's regular activities by:
  Planning meetings
  Preparing meeting agendas
  Conducting meetings
  Following up on PIT activities
  Resolving or escalating differences between PIT members
- Safeguard the integrity of measurement data
- Identify critical success factors and key dependencies of the process
- Define subprocesses and their owners (usually line managers)
- Direct the various BPI stages
- Prepare documents of understanding
- Monitor process qualification and process benchmarking activities
- Identify and implement process changes required to meet business and customer needs
- Maintain contact with the BPI champion and the EIT regarding:
  The PIT's progress
  Process qualification
  Requests for special investments
  Automation and mechanization issues
- Overcome obstacles to improvement by:
  Ensuring that proper resources are available to the PIT
  Settling interdepartmental conflicts
- Establish the appropriate mechanisms for continuously updating procedures and improving the effectiveness and efficiency of the overall process
- Maintain contact with the customers of the process to ensure that their expectations are understood and met

- Keep the PIT informed about business changes that may affect the process

Other Concerns. When a company begins the improvement process, middle and top managers often voice a concern about time. Despite their commitment to improvement, some may fear their people will be overextended as team members or leaders. Most process owners will want to know the time requirement in advance. Although no two situations are identical, the process owner should expect to devote a significant amount of his or her work time to BPI (i.e., 20 hours or more a week) during the first month of the job. After this initial period, a few hours per week to prepare for, attend, and follow up on meetings will be necessary.

This time commitment could last for a few months or a few years, depending on the pace of improvement and the extent of the change required. After reaching the process qualification goal, the process owner should devote regular, but shorter, periods of time to process monitoring (e.g., tracking the effectiveness and efficiency measures, and comparing the process to world-class standards). However, as specific problems arise, or as benchmarking with other companies uncovers interesting areas for improvement, some intensive periods of work will be required. The process owner's performance plan should reflect these new responsibilities, and his or her rewards should be based on how well the total process operates.

PIT Members

General Scope of the Job. A PIT member's primary responsibility is to represent his or her department on the team.

Major Responsibilities. The job description should include the following responsibilities:

- Participate in all PIT activities (e.g., train in BPI techniques, attend meetings, and walk through activities)
- Conduct BPI activities in his or her department as required by the PIT (e.g., obtain "local" documentation, develop a flowchart of the department's participation in the process, verify application of the process, measure efficiency, and help implement departmental changes)
- Obtain appropriate resources (e.g., time) for the activities to be performed within his or her department
- Implement changes in the process as they apply to his or her de-

partment (e.g., supervise production of new documentation, organize training, and perform follow-up work)
- Chair subprocess teams as appropriate
- Support change (e.g., inform, encourage, provide feedback, and listen to complaints)
- Train and involve other department members as appropriate
- Solve process-related problems
- Provide his or her department with a better understanding of how it fits in the total process

Other Concerns. PIT members' concerns may resemble those of the process owners. However, the time requirement differs. PIT members should expect to devote roughly 4 hours per week. Depending on the approach selected, the time required could be as high as 50 percent during the first 2 months. For example, the EIT might decide that a critical process (e.g., billing, accounts payable, or shipping) requires immediate attention. This could mean a full-time assignment of about 4 weeks for team members. The full-time approach is frequently used to prepare a business process for computerization or to make a significant immediate improvement. Although the EIT should be ready to make the appropriate investments to reach its goals, launching a PIT generally will not require this level of resource commitment.

GENERAL MANAGEMENT RESPONSIBILITY

We have talked a lot about PIT responsibilities. Now let's address management's responsibilities to BPI. Management support to BPI at all levels is critical to the success of the process. It will require the active involvement of all the management team involved in the selected processes, as well as the managers who provide input into these processes and receive output from them. General management's responsibilities are:

- Providing the resources required, including staffing and capital
- Developing common objectives that support the proposed changes
- Breaking down walls that stand between organizations
- Searching out improvement opportunities
- Setting up department improvement teams to support the processes being evaluated
- Changing its own thinking to get a total process perspective
- Providing the necessary training and education to support the new processes

- Anticipating the impact of process changes on their organization and preparing for them
- Establishing systems and reviews to ensure that the progress does not degrade
- Rewarding teams and individuals who make significant contributions to BPI
- Showing interest in BPI by frequent reviews of status and results
- Finding equivalent or better jobs for people whose jobs have been eliminated as a result of BPI

That last responsibility warrants more discussion because it is so important. Management cannot expect people to evaluate the business processes fairly and look for ways to improve them if it means that they or the person working beside them will be laid off. We believe that management should develop a no-layoff policy that would read something like this:

> We will not lay off anyone as a result of an improvement idea or a process change resulting from a PIT suggestion. When jobs are eliminated, the individuals will be reassigned to an equivalent or better job. This does not mean that we will not lay off people because of a downturn in business.

Without this type of assurance, management cannot expect the full cooperation of the members of the PIT or their management. The old barriers between departments will stay in place. People will hide waste to protect themselves, their friends, and their employees. Effectiveness may improve, but you will not develop the most efficient process.

We know of one organization in which improvement activities eliminated more than 100 jobs. Attrition took care of about 50 percent of the jobs. There were still more than 50 employees whose jobs were eliminated. To handle this problem, a contest was held to select 50 employees who would go to school at a local university until jobs were available. The company paid the employees' salaries and tuition while they were in school. As a result, the company has more improvement ideas than it can handle. All employees are looking for ways to eliminate their jobs so that they can go back to school.

SUMMARY

It is very important to understand any undertaking; this is particularly true when the undertaking will revamp the way you manage your business. Don't rush into the improvement stage until your plan is well defined and evaluated. Then be sure that you match the right person to the right job. Taking the time up front will save you many false starts and, in the long run, will save you time and money. This phase of the BPI cycle should not take more than 2 months. Let's take the time to do BPI right the first time, because if we don't, our employees and our customers may not give us a second chance.

ADDITIONAL READING

Harrington, H. James, *Poor-Quality Cost*, ASQC Quality Press, Milwaukee, WI, and Marcel Dekker, New York, 1987.

_____, *The Quality/Profit Connection*, ASQC Quality Press, Milwaukee, WI, 1989.

3

Organizing for Process Improvement

INTRODUCTION

The complexity of most business processes makes it necessary to formally organize improvement activities. A team approach to business process improvement (BPI) provides long-lasting results and minimizes implementation time. This chapter explains how to form and organize a typical process improvement team (PIT), define the process boundaries, and start the process measurement system.

PROCESS OWNER EDUCATION

In most cases, training in their new role and in BPI methods should be provided to process owners as soon as they are selected. A 1-day overview will serve this purpose and allow the new process owner to function with a great deal of confidence. The process owner and the team members will receive much more detailed training after the team has been assigned.

INITIAL PIT PLAN

Once the process owner has completed the introductory class on BPI, he or she will prepare a PIT plan that will define the activities and time schedule up to the point at which the final boundaries are established. Some of the activities that will be addressed in this plan are:

- Defining preliminary boundaries
- Block diagraming the process
- Updating operating assumptions
- Interviewing managers involved in the process
- Forming the PIT
- Providing initial team education
- Developing the PIT mission statement
- Setting the final process boundaries

The process owner can partly control the time to complete this part of the process. Many activities in this phase of the process require the co-operation of management from other organizations. Getting their buy-in and their commitment to provide resources is always time consuming. Our experience indicates that most estimates of the time for this phase of the process are conservative and need to be revised later on. The exception to this is when each EIT member has personally been very active in communicating the importance of the BPI concept.

PRELIMINARY PROCESS BOUNDARIES

The process owner's first job is to define the beginning and end boundaries of the preliminary process (i.e., where the process begins and ends). The modern business organization comprises a complex maze of interactive, interconnected, and/or sequential processes. Defining the process boundaries to break this maze into logical, manageable pieces is a crucial task.

In most processes, the beginning and end points are not clearly defined. One person might view them in a very limited sense; another, in a more global manner. As a result, the owner's task of identifying them is not as easy as it sounds. Consider, for example, a familiar activity like preparing a barbecue steak dinner. Based on the process statement, there are many points at which the process could begin. Some of the beginning points could be:

- Getting a job so that you have money to buy the steaks
- Deciding to have a steak barbecue
- Going to the store to buy the food
- Taking the steaks out of the refrigerator to cook them

The process also has many potential end points. They could be:

- When the steaks are cooked
- When the steaks are served
- When the steaks are consumed
- When the mess is cleaned up and put away

As you can see, there are many potential combinations of beginning and end boundaries. Moreover, in addition to these boundaries, you must establish upper and lower boundaries to limit the complexity of the process and to clearly define basic assumptions. Adding the upper and lower boundaries to the beginning and end boundaries literally boxes the process in.

In the barbecue example, the upper and lower boundaries might be that we would not include the process of preparing the grill, furnishing dinner plates, inviting the guests, or making the potato salad. These all will be considered inputs to the process because they represent activities outside the scope of the core process. Figure 3.1 demonstrates how the boundaries can be used to box in the steak barbecue process.

Both the beginning and the upper boundaries allow inputs to enter the process. All inputs to the first activity in a process enter through the beginning boundary. The upper boundary allows inputs to enter any other activity in the process. The lower boundary allows output from the process to go to secondary customers at any point in the process, while the output from the end boundary is the primary output from the process and goes to the end customer of the process. The selection of these boundaries determines who will be involved in the process and what goes on within it. In Figure 3.2, point A is another point that could have been chosen as the end boundary.

BLOCK DIAGRAMING THE PROCESS

To help identify the key departments in the business process under evaluation, the process owner should make a block diagram of the process and identify who is performing the key operations. Chapter 4 provides details on block diagraming. Figure 3.3 presents a typical block diagram for a summer barbecue, with the responsible people listed for each operation.

This is a very important step in the process since it forces the process owner to mentally walk through the total process. It is strongly recommended that the process owner do some research before he or she starts to construct the block diagram. He or she should read the relevant procedures and talk with people in the process. The process owner must have a good view of the total process before putting pencil to paper.

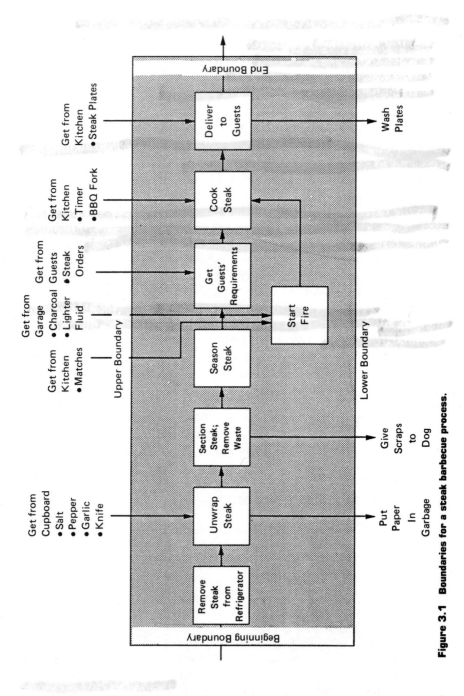

Figure 3.1 Boundaries for a steak barbecue process.

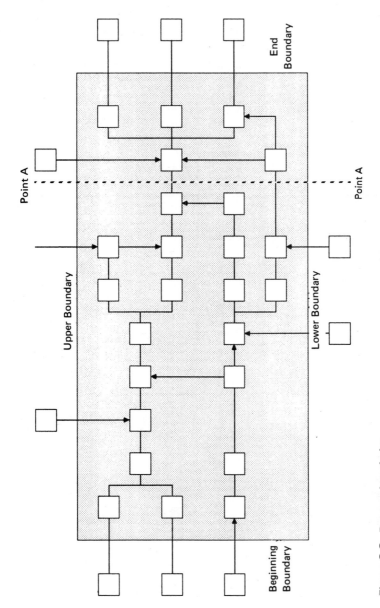

Figure 3.2 Process boundaries.

59

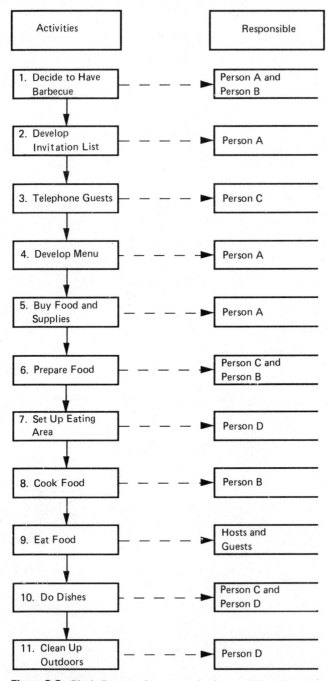

Figure 3.3 Block diagram of a summer barbecue with assignment of responsibilities.

This is a great exercise for the process owner, who will be surprised by how much can be learned from constructing this simple block diagram.

UPDATING THE OPERATING ASSUMPTIONS

The process owner will now need to update and expand on the original operating assumptions developed by the EIT. These assumptions include how frequently the team will meet and the length of the meeting, as well as many other operational details. These details should be defined so that they can be communicated to the managers involved in the process. These operating assumptions, plus the process block diagram, document how and where the PIT will be directing its activities. Typical items that would be included in the updated operating assumptions are:

- Employee resources required
- Project end date
- Capital equipment limitation
- Any committed changes to process inputs
- Anticipated changes to customer expectations
- Areas in the process that will not be changed
- Services to be provided by areas not represented on the PIT

The operating assumption list is an active document that can and should be changed as new information becomes available. At a minimum, the operating assumption list should be reevaluated after the process owner has interviewed all other managers involved in the process and after a flowchart of the process has been made.

TEAM STRUCTURES

Many types of teams become involved in the BPI activities. We have already talked about the executive improvement team (EIT) and its role in the process. Other types of teams that can be used are:

- Process improvement team (PIT)
- Subprocess improvement team (sub-PIT)
- Task team (TT)
- Department improvement team (DIT)

Figure 3.4 shows a typical organization chart and how each type of team fits into the organization.

The PITs and the sub-PITs are the most frequently used teams, and the rest of this chapter will describe their organization and functions in great detail. DITs are often in place before BPI methods are applied.

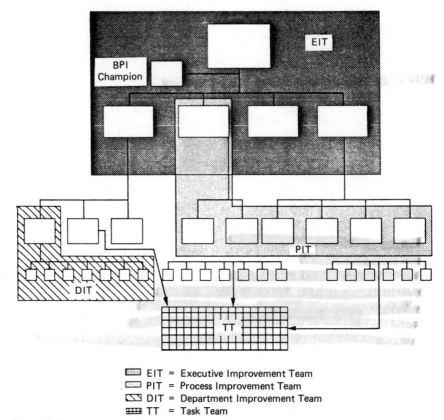

EIT = Executive Improvement Team
PIT = Process Improvement Team
DIT = Department Improvement Team
TT = Task Team

Figure 3.4 Typical team structures.

This is the way all members of the organization should be introduced to the team and the problem-solving tools. TTs are organized after initial work has been performed by the PITs. Information regarding their organization and functions is also provided in this chapter.

Process Improvement Team

PIT is a most appropriate acronym for the process improvement team. The pit is the center of many of the delicious fruits we eat. It is the seed that brings about new life, new growth, and increased productivity. The PIT also is the center of our improvement activity. Its efforts will bring about a new way of thinking about our business and the way our process functions. As with the pit in a peach or a plum, proper nurturing of the PITs will bring about new growth for your organization and increased effectiveness, efficiency, and profits.

A PIT should include representatives from each department involved in the process. The department managers should assign team members who will be responsible for making commitments for the entire department.

A PIT normally will consist of 4 to 12 members; more than 16 members will reduce its effectiveness. The PIT will be responsible for designing and continuously improving its assigned process. Typical team activities include:

- Flowcharting the process
- Gathering process cost and quality information
- Establishing measurement points and feedback loops
- Qualifying the process
- Developing and implementing improvement plans
- Reporting efficiency, effectiveness, and change status
- Ensuring process adaptability

PIT activities usually will be heaviest during the first 1 to 3 months. Between 25 and 50 percent of the members' time will be required to define, verify, and update the process and procedures. After that, the time requirement may be limited to an hour-long meeting every 1 to 2 weeks. Additionally, PIT members will be working with other people within their areas to improve their sections of the process. During this stage, PIT meetings will be devoted to change status reviews, with members reporting on individual activities.

Subprocess Improvement Team

When a very complex process is being improved, it may involve many areas, making it impractical for representatives from all the affected areas to be members of the PIT. In other cases, it may be advantageous to have a small group work on different sections of the total process. In these cases, it is very effective to divide the macroprocess into subprocesses and have a subprocess owner and sub-PIT assigned. The subprocess owner is always a member of the PIT, and he or she provides a progress report at the regularly scheduled PIT meeting.

Task Team

Frequently, a PIT will identify a major interfunctional problem related to the business process the PIT members are trying to improve. When this occurs, a TT should be assigned to solve the problem. Similarly, when major changes to the process are required (e.g., automation), a dedicated group of experts should be assigned to a TT to ensure that the change is implemented correctly. The process owner, with the approval of the participants' managers, should assign the job. TT members' responsibilities will continue until the problem is solved or the process change has been implemented and its effectiveness and efficiency measured.

Department Improvement Team

DITs comprise all the members of a particular department or small work group (supervisor or first-line manager and all the employees that report to him or her). They provide a focus and a means for all employees to contribute to an ongoing activity aimed at improving the effectiveness and efficiency of their department. The department manager often serves as the team's chairperson, although this activity may be performed by a trained and capable nonmanagement employee.

A DIT identifies problems that cause errors and/or conditions that decrease the department's effectiveness. It then develops and implements corrective actions to eliminate these roadblocks to high productivity and/or error-free performance.

The team's efforts focus on activities within the department or on those that directly impact departmental activities. The team defines problems, sets priorities, selects improvement targets, and implements activities that will enable the department to meet or exceed these goals.

DIT members can make very important contributions to the improvement process and should be encouraged by the PIT to become involved in their own improvement activities. This develops the company's human assets and reinforces the feeling that the improvement initiative is a total and integrated process. BPI is designed so that a member of the PIT is also a member of the DIT for each department involved with the process being improved. This provides a close working relationship between the two types of teams. If the DIT cannot solve a problem because it requires action from another department, the PIT can often get it corrected. As a rule, most of the problem solving and improvement that relate to the department are done by the DIT, leaving the PIT to serve more as program management. These various groups develop gradually. The BPI effort will gain momentum as the process owner and PIT members make key assignments.

SELECTING PIT MEMBERS

After defining the preliminary boundaries and block diagraming the process, the process owner should determine which departments play key roles in the process. Each of these departments should be represented on the PIT. These representatives will communicate and coordinate activities between the PIT and the DITs or department managers.

The PIT member will facilitate implementation of the necessary changes to the department's process.

To ensure that the correct people are assigned to the PIT, and to obtain an initial assessment of the process, the process owner should meet with the manager of each key department to discuss the following items:

- The PIT's purpose
- The PIT's objectives
- The amount of involvement that its organization has in the process
- PIT members' responsibilities
- Key process supplier on whom the manager depends
- To whom the PIT provides process output
- Problems the manager is experiencing with the process
- Improvement suggestions for the process
- The names of department representatives

Do not underestimate the importance of these meetings. If well managed, they will provide the process owner with important information about the process that he or she will find extremely valuable as the improvement efforts progress. They may even lead to new process boundary definitions. Most companies have never really studied their business processes, so it usually is difficult to initially determine their exact focus and scope. These meetings should be seen as a means of furthering the business process definition.

The process owner should point out to the manager that the department's representative to the PIT should be an "expert" in his or her knowledge and understanding of the detailed activities performed in that area of the process. It is extremely important that the very best person be assigned. The process owner should be sure that the department manager understands the role the department representative plays and the amount of time required. It is a good idea to give him or her a copy of the job description for the PIT member. In many cases, because of time limitations, lack of detailed knowledge of the process, or because another department member can perform the assigned tasks better, the manager will select one of the employees as the PIT representative. The selected department representative should have:

- The authority to commit the department resources
- The time to participate on the PIT
- The time to follow up on assignments given during the PIT meeting
- Practical and actual process knowledge
- Credibility with the other PIT members
- A desire to be part of the BPI activities
- A belief that the process can improve

- The willingness to embrace and lead change
- A vested interest in the process

If for some reason the department cannot assign a representative, the issue should be escalated immediately to the EIT or the BPI champion. In many cases, a customer of the process is also a member of the PIT. This helps keep the PIT focused on the customer. Frequently, the process suppliers are also members of the PIT and play a very important role. We have found that the most effective teams include representatives of their suppliers and customers as part-time members.

PIT ORIENTATION

Now that the PIT members have been identified, we need to prepare them for this important assignment. Prior to the first meeting, the process owner should send the PIT members the following:

- The process goals and assumptions
- The process block diagram
- A list of the PIT members' addresses and phone numbers
- Copies of this book, with instructions to read Chapters 1, 2, and 3
- Copies of the BPI directive
- Copies of the PIT member job description
- The agenda for the first meeting and training session

We have found it effective to bring together two PITs for their first meeting and training session. This session provides an effective way of introducing the PITs to the BPI technology while working on their assigned processes. This jump start will provide the team with the needed understanding of the BPI methodology and help them develop their mission statement and end boundaries.

TEAM TRAINING

PIT members must be trained to work as a team, understand the process, collect and analyze data, and improve the process. As a prerequisite to becoming a PIT member, each individual should have been trained in and should have used basic team and problem-solving tools such as:

- Team process
- Brainstorming
- Check sheets
- Graphs

- Histograms (frequency distributions)
- Pareto diagrams
- Scatter diagrams
- Nominal group techniques
- Delphi narrowing technique
- Force-field analysis
- Cause-and-effect diagrams
- Mind maps
- Statistical process control

If members have not been trained in the basic team and problem-solving tools, the PIT should begin with a training class on this subject. Lack of training always has long-term negative results. At the start of team efforts, the level of enthusiasm usually is so high that groups will jump right in and may even achieve some results. In the long run, however, a team lacking training and skills will not completely comprehend the situation it is trying to improve and will not implement the best combination of solutions.

John Young, president of Hewlett Packard, noted:

> It became quite clear after we began operations that many of the participants did not have the conceptual and personal skills needed to make them effective. The same perception became more clear as we moved to just-in-time manufacturing. It lays bare the bones of group activities. There was simply no slack to cover up a team's glitches or weak points.

BPI's 10 Fundamental Tools

In addition to such basic team dynamics and problem-solving training, the PIT should have some specialized training to prepare its members for the assigned activities. This training should include, but not be limited to, the following:

- BPI concepts
- Flowcharting
- Interviewing techniques
- BPI measurement methods (cost, cycle time, efficiency, effectiveness, adaptability)
- No-value-added activity elimination methods
- Bureaucracy elimination methods
- Process and paperwork simplification techniques
- Simple language analysis and methods
- Process walk-through methods
- Cost and cycle time analysis

These 10 fundamental tools will be discussed later in the book in greater detail.

BPI's 10 Sophisticated Tools

As BPI activities advance, the PIT may feel the need for more sophisticated tools to reach still higher goals. Team members should then increase their abilities by learning and using some of the more sophisticated tools. The tools are:

- Quality function deployment (QFD)
- Program evaluation and review technique (PERT) charting
- Business systems planning (BSP)
- Process analysis technique (PAT)
- Structured analysis/design (SA/SD)
- Value analysis
- Value control
- Information engineering
- Benchmarking
- Poor-quality cost

For further information on these sophisticated tools, contact the author.

UNDERSTANDING THE ASSIGNMENT

For any improvement effort to succeed, its mission and scope must be clearly stated and understood. For a process improvement activity, this involves understanding and/or defining:

- BPI objectives provided by the EIT
- Operating assumptions
- Preliminary process boundaries (already completed by the process owner)
- Process mission statement
- PIT name
- Final process boundaries

Mission Statement

One of the first activities the PIT should undertake is to prepare a mission statement clearly defining its assignment. For example, a mission statement for a design release PIT might be:

> To understand and apply BPI methods to the total design release process to make it more effective, efficient, and readily adaptable to ever-changing

business needs. The results will include reducing the costs to release a new design, reducing the total cycle time, and improving the manufacturability of the design. The new process will be completely documented and agreed to by the organizations involved, and should be in use by January 16, 19—.

A good mission statement should:

- Be short (never more than five sentences)
- Define the scope of the activities
- State what will be accomplished
- In some cases, include performance improvement targets and completion dates

Most PITs find that it is helpful to develop a simple two- or three-line description of the goals and/or objectives of the process under study before they prepare the PIT mission statement.

While it is appropriate for the process owner to suggest what he or she thinks the mission statement should look like, great care should be taken to ensure that the PIT, as a whole, buys into it. Each word should mean something special to the team, and all members of the team should have exactly the same understanding of what needs to be accomplished. If prepared too rapidly, or without adequate ownership, the mission statement could become a meaningless document, when it should provide clear direction and become the measure of success for the PIT.

As soon as the PIT develops a mission statement, it should also develop a name for the team. The PIT name should reflect the mission statement and include the names of the products and services provided by the process.

FINAL PROCESS BOUNDARIES

Given the mission statement, the PIT's second task should be to reevaluate the preliminary boundaries determined by the process owner to see whether they are correct or whether they need to be adjusted and/or better defined. The boundaries should be compared with the activities specified in the mission statement, to ensure that they and the mission statement agree. If the PIT decides on process boundaries different from the preliminary boundaries, it may be necessary to adjust the PIT membership in keeping with the new boundaries.

The process boundaries will define the following:

- What is included in the process
- What is not included
- What the outputs are from the process

- What the inputs are to the process
- What departments are involved in the process

Be sure that the boundaries are wide enough to cover known problems and that the scope of the process is in the best interest of the customers.

The question of reducing process size is a delicate one. For example, it may seem advisable to divide the process in Figure 3.2 at point *A*, creating two separate processes. While processes must be of a manageable size, the danger is that such reductions tend to cause suboptimization unless there is a logical breaking point.

The best approach to the problem of process size is to create sub-PITs that contain a number of logically connected operations. An effective way to organize a sub-PIT is to have it focus on the activities that are contained within a number of departments from the same function. The leaders of the sub-PITs should be members of the original PIT. Take the example of a proposal process for a large construction company that involves a dozen departments and can take as long as 6 months. An effective way to study the macroprocess would be to break it down into subprocesses, such as proposal review, engineering design, cost estimation, and pricing, and then create sub-PITs assigned to each subprocess.

Revised Block Diagram

Once the process boundaries have been determined, the PIT should direct its attention to the block diagram originally prepared by the process owner. Each PIT member must understand and agree with this block diagram. In most cases, the original block diagram will be extensively revised during this review, with more inputs and outputs added. The end result of this cycle should be a block diagram that is boxed in, understood, and supported by the total PIT. This microprocess block diagram will become the heart of the PIT activities. It should always be kept up to date. Revise it whenever new information is obtained and as process changes are implemented.

PROCESS OVERVIEW

We now have a PIT that has an understanding of process boundaries and has also created a block diagram of the process. There are other essential pieces of information that should be gathered and understood before we can plunge into any detailed analysis. We need to understand:

- Who the suppliers of the inputs to the process are
- Who the customers of the outputs of the process are
- What other processes it interacts with

Typically, inputs or outputs of business processes are information or services (e.g., data, documents, reports). Time should be spent identifying all significant inputs and outputs. Processes typically have several different inputs and outputs, although only one or two may be considered as primary inputs and outputs and the rest, secondary. For example, for a cash collection process:

- Primary input would be invoicing information.
- Secondary inputs would be accounts receivables data, supporting documents, etc.
- Primary output would be collected invoices.
- Secondary outputs would be summary reports, credit notes, etc.

The inputs and outputs should be reviewed to ensure that there are no inconsistencies or gross omissions. Once they have been defined, the suppliers and customers can be identified. Suppliers are typically other processes or departments (and in some cases, external suppliers) that provide the input. There may be several suppliers for the same input. Using the previous example, the shipping process would be the supplier of invoicing information for goods shipped, and the training process would be the supplier of invoicing information for training provided.

Similarly, each output may have one, or several, customers. It is especially important to identify customers, for several reasons. First, it is critical in setting measurements for the process. Second, people in most organizations are unaware of who their customers are, so this identification is frequently an eye-opener. To identify customers, you need to find out who receives or benefits from the output of the process. That department, process, or organization (internal or external) is the customer. Using the example of the cash collection process again, the customers for collected invoices would be all the divisions for which this function is performed, and the customer for summary reports would be upper management.

Finally, it would help the PIT to understand the interfaces that the process under review has with other processes so that it understands the "big picture" and coordinates its efforts with other PITs, if necessary.

To accomplish this, the PIT should develop a list of all inputs to, and outputs from, the block diagram. Each input and output should be classified as primary or secondary. Be sure that all primary outputs and their customers are identified. The secondary outputs and all inputs will be further refined when the process is flowcharted. For the time being, this list will be used to ensure that the boundaries are correct and to define the end customer.

In many cases, primary customers are invited to join the PIT. This close working relationship helps build a supplier-customer partnership. Experience has shown that when primary customers are included on the PIT, there is a much better return on investment because they are in the best position to make trade-offs. Often, they can make a slight change in their process that results in a major efficiency improvement in the process under study. Without this close interface, the customer has little understanding of the work involved in meeting his or her minor needs and is therefore reluctant to relax requirements.

PROCESS CUSTOMERS

Who are the customers of the business processes? How do you identify them? Anyone (person or organization) who receives output from the process, either directly or indirectly, is a customer. Customers can be within the organization (internal customers), outside the organization (external customers), or both.

A single process can have as many as five different types of customers (see Figure 3.5). They are:

1. *Primary customers.* They are the customers who directly receive the output from the process. The primary customer for the output from the parts order computer run would be purchasing.

2. *Secondary customers.* A secondary customer is an organization outside the process boundaries that receives output from the process but is not directly needed to support the primary mission of the process. Often, the output that goes to a secondary customer outside the process

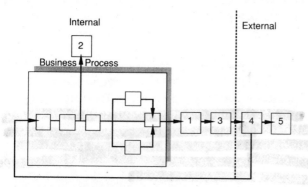

Figure 3.5   Order entry customer profile.

is also used within the process. The secondary output is needed to trigger other business processes; therefore, these outputs are important, although they may or may not contribute to the primary mission of the process under study. An example of a secondary output would be: When a computer run is completed, the computer operator removes the tape, signaling to another operator that the computer is now free to run another job.

3. *Indirect customers.* These are customers within the organization who do not directly receive the output from the process but are affected if the output from the process is wrong and/or late. Manufacturing is an indirect customer of the parts order computer run process because if the computer run is not right, manufacturing will not be able to meet its ship schedule.

4. *External customers.* These are the customers outside the company who receive the end product or service. An example of an external customer is a car dealer who is an independent businessperson franchised by one of the auto manufacturers to sell its products.

5. *Consumers.* These are often indirect, external customers. Sometimes, companies deliver their output directly to the consumer. In these cases, the external customer and the consumer are the same person or organization. In most cases, products and/or services are delivered to a distributor, a representative, or a store that sells the product and/or services to the consumer.

Figure 3.5 is a customer profile of the order entry process. The end output from the process goes to customer 1, the primary customer. In this case, the output is the parts order form that goes to purchasing. Earlier in the process, a secondary output goes to customer 2, the secondary customer. In this case, it is a report generated for corporate headquarters that is used to balance work loads between manufacturing plants. Customer 3 is an indirect customer. In this case, it would be manufacturing, which would be receiving the parts to manufacture a television set. Its output goes to customer 4, who is outside the corporation and could be a major distributor, like Sears. This is the external customer for the corporation that is producing the television set. Customer 5 is the consumer who buys and uses the television. For some processes, the primary customer is the consumer. No matter what the process is, all five types of customers are important.

Throughout this book, when we use the word *customer,* we will be focusing on the primary and secondary customers, but all five types should be considered in improving your process. To help you identify your customers, ask your team:

- Who receives the output from the process?
- What do they expect from the process?

- How do they use the output?
- What impact does it have on them if it is wrong?
- How do they provide you with feedback if it is wrong?
- How far beyond the primary customer will errors in the process output impact the effectiveness and efficiency of the operation?
- Are there other outputs that are generated during the process?

The PIT should make a list of all the process customers and get to know as many as possible on a one-on-one basis.

EFFECTIVENESS, EFFICIENCY, AND ADAPTABILITY MEASUREMENTS AND TARGETS

Let's review what you should have accomplished so far. You have a team. You have a mission statement. You have boundaries. Now, what should the PIT improve? This is where the issue of process measurements and targets comes in.

What is the goal of our effort? What would we like the streamlined process to achieve? Setting targets ensures that you focus on what you want to improve. Therefore, the next step is to establish measurements and targets for the overall process. We purposely speak of measuring the *overall process* because at the start of the improvement process you should not waste energy on details that may prove unnecessary once the process is streamlined.

In establishing measurements and targets for outputs from all activities, you may find that the number becomes very large and unwieldy. It is therefore important to limit the requirements to a critical few for the total process.

There are three major process measurements:

- *Effectiveness.* The extent to which the outputs of the process or subprocess meet the needs and expectations of its customers. A synonym for effectiveness is quality. Effectiveness is having the right output at the right place, at the right time, at the right price. Effectiveness impacts the customer.
- *Efficiency.* The extent to which resources are minimized and waste is eliminated in the pursuit of effectiveness. Productivity is a measure of efficiency.
- *Adaptability.* The flexibility of the process to handle future, changing customer expectations and today's individual, special customer requests. It is managing the process to meet today's special needs and future requirements. Adaptability is an area largely ignored, but it is critical for gaining a competitive edge in the marketplace. Customers always remember how you handled, or didn't handle, their special needs.

Let's discuss these requirements in detail, since they form the basis for improving the process.

Effectiveness Measurements

To ensure that the process is effective, you must define the customer needs and expectations and then meet those needs and expectations. The first step should be to determine what your customer needs and expectations are. The second step should be to specifically describe those needs and expectations in measurable terms. The third step is to define the way the measurement data are collected and used.

Customer needs and expectations typically relate to products and/or services:

- Appearance
- Timeliness
- Accuracy
- Performance
- Reliability
- Usability
- Serviceability
- Durability
- Costs
- Responsiveness
- Adaptability
- Dependability

The PIT should meet with the primary customers to determine what they require from the process. The customers always know what they want ("voice of the customer"), but frequently, they have a hard time expressing their desires in terms that can be used to measure effectiveness. Often, it is heard in general terms like "I want quick service," "error-free reports," "things that work," or "easy-to-use output."

What the process needs is measurable characteristics that can be:

- Evaluated before the output is delivered to the customer
- Documented in a specification so the employees have a standard
- Agreed to by both the supplier and the customer

It is not always possible to adequately document customer needs and expectations with words. Sometimes examples, samples, and pictures are required to adequately demonstrate the criteria.

We have talked a lot about needs and expectations. They are different, and you should know the difference. Needs set the minimal standard for the process output that the customer will accept. Ask yourself what you *need* in a hotel room. You need a bed, a place to wash, a place to hang up your clothes, and temperature control. Needs are normally reflected in engineering specifications. Now ask yourself what you *expect* in a hotel room. Probably you would add a TV, a desk, shampoo, etc. As you can see, there is a difference between needs and expectations. Sometimes, needs set the higher standard. In these cases, our customers screen the process output to ensure that it is correct. Multiple signatures on a purchase order are an example. In most cases, though, our expectations set the higher standard. Needs normally don't change, but expectations change very often. Because of this, we must stay very close to our customer and keep an open communication channel so that we can keep the output from our process better than our customers' expectations.

The PIT should work with the primary customers of the process to convert the voice of the customer into effectiveness measurements and targets that both the PIT and the customer agree to. It is best to put these agreements in writing, signed by both parties. In addition, the PIT must develop a way to measure compliance to the criteria. Some techniques used to measure compliance to customer expectations are:

- Check sheets filled in by the customer and returned to the supplier
- Feedback of customer sampling of incoming products and/or services
- Self-inspection as the customer views the product or service
- Surveys and/or questionnaires
- Focus groups
- Interviews with customers
- Monitoring customer complaints
- Market research

Effectiveness measurements must use input from both external and internal customers. Although some processes interface directly with the external customers, all processes have internal customers and, through a chain of transactions, provide an output to the external customer (see Figure 3.6).

Hence, ignoring external customer expectations for any process being studied would lead to less-than-desired improvements. A good example is the external customer's expectation for on-time delivery. This is not the sole responsibility of the shipping process, which interfaces with the external customer. In a simplified situation in which parts are normally shipped from stock, the order entry process provides customer order information to the customer service process, which provides scheduling information to the material storage process, which pulls parts and sends them to the shipping process, which ships them to the external customer.

Therefore, if we are studying the order entry process, we must un-

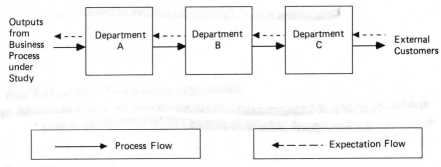

Figure 3.6   Deploying external customer requirements.

derstand the indirect impact of external customer requirements on that process. This would enable the order entry process to become more customer focused, which is what all world-class organizations must strive for in all their processes. Deploying customer needs and expectations deep into the organization would help all employees understand the ultimate impact of their work on the external customer. Continuing with the example of the order entry process, it would enable an order entry clerk to understand how prompt and accurate data entry ultimately affects on-time shipments to the customer. This understanding also brings pride and meaning to jobs that might otherwise appear inconsequential.

The PIT should take the primary outputs from the process under study and block diagram the sequence of events triggered by those outputs that have an impact on the final external customer. This will help the PIT understand each area's requirements in order to fulfill external customer needs and expectations.

We have discussed the techniques for understanding external customer requirements and evaluating their importance. We now need to understand how to convert these requirements into measurements for the process we are studying. To accomplish this activity, the PIT should:

- Quantify the external customer needs and expectations
- Understand the chain of processes that fulfill those needs and expectations
- Set targets for each process so that customer needs and expectations are met

It is conceivable that the PIT may need to interface with other PITs in developing these measurements. Additionally, the EIT should manage any issues that are outside the authority of the PIT.

After gathering this information, the PIT should determine the extent to which actual results conform to customer expectations. A minimum of one effectiveness measurement should be established for every primary customer.

Installing measurable requirements (needs and expectations) not only ensures conformance but provides a means of quickly identifying the cause of an error, if one occurs. Customer expectations must be detailed enough that any person, even one not familiar with the process, can judge whether or not the needs and expectations are being met.

Efficiency Measurements

In addition to effectiveness, efficiency is also important to the external customer. All organizations must consistently work to make all their business processes more efficient, because as operating costs are reduced, some of the savings should be passed on to the external customer.

Lack of effectiveness is easy to see and measure. Poor efficiency, on the other hand, is harder to recognize. We learn to live with it, and we slowly let it get worse. We add more checks and balances when we have a problem, or more people are added to the process, and we never remove them. An organization must minimize the resources required to do each job and eliminate waste and no-value-added activities, to achieve efficiency. Requirements for efficiency focus on the use of money, time, and other resources. Typical efficiency measurements are:

- Processing time
- Resources expended per unit of output
- Value-added cost per unit of output
- Percentage of value-added time
- Poor-quality cost
- Wait time per unit

Among the efficiency requirements, one of the most meaningful usually is cycle time. For most business processes, the cycle time is unduly long. For example, in hiring a person, the delay between the need identification and hiring often can be calculated in terms of weeks or months, although the processing time is less than 36 hours in most cases. Even a rough measurement of cycle time usually will enable PIT members to discover the no-value-added or no-service-added activities existing within the process. Therefore, although our first goal is to meet customer needs and expectations, efficiency improvements may come first in order to free up resources required to implement the improvements needed to meet customer needs and expectations.

Another measure of efficiency is output versus input (e.g., miles per gallon). Accordingly, one of the key elements to measure will be the value-added time versus no-value-added time. Typically, the real-value-added time (e.g., the time required to take an input and transform it with no wait, transportation, and/or checks and balance time) is only 5 percent of the total cycle time. Unfortunately, instead of working on the 95 percent no-value-added time, we spend most of our time speeding up the real-value-added activities. Our focus must be expanded.

Many organizations try to have the real-value-added time be no less than 50 percent of the total cycle time. This is a good start but not world-class. A target of 90 percent is an obtainable goal. Consequently, all the work you do to streamline the process will help minimize delay time.

Efficiency can also impact our external customers by not allowing the organization to deliver output on schedule. We need to relate external customer expectations to the business process in the same way we did to establish effectiveness measurements and requirements. Figure 3.7 shows graphically how this can be done.

In this example, the customer expects shipment of parts within 72 hours of ordering them. This is achievable only if order entry performs its function (checks, batches, inputs, edits, and transmits orders to customer service) within 8 hours of receiving the order. This then becomes an effectiveness measure for the total order entry process.

Improved efficiency also requires eliminating the errors that occur in the process. Error-free performance of process tasks by every individual involved in the business process must be our objective. It means that we must develop processes that will not allow errors to occur, and if errors do occur, we must react quickly to prevent them from recurring. Each process and subprocess should have a minimum of two efficiency requirements established: one that is a ratio of input-to-output value, or resources per unit output (for example, hours per change, cost per new employee, cost per purchase requisition processed) and another that measures cycle time (for example, hours to process a change, days to hire an employee, hours to process a purchase order).

Adaptability Measurements

Basic quality involves satisfying customer requirements. However, for many consumers today, *Good enough is not good enough any more*. Companies must exceed customer needs and expectations, now and in the future, by:

- Empowering people to take special action
- Moving from meeting basic requirements to exceeding expectations

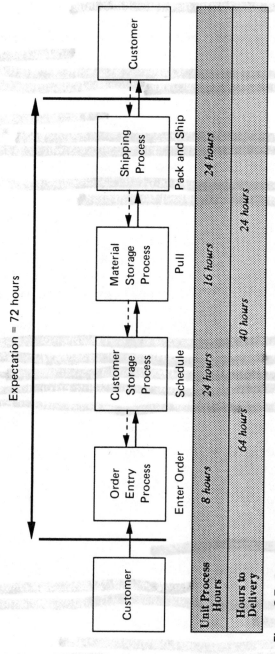

Customer Expectation = *On-time* shipment
Quantified Expectation = Shipment *within 72 hours* of order

Expectation = 72 hours

Unit Process Hours	8 hours	24 hours	16 hours	24 hours
Hours to Delivery	64 hours	40 hours	24 hours	

Enter Order — Schedule — Pull — Pack and Ship

Figure 3.7 Relating external customer expectations to the business process.

- Adjusting and adapting to ever-changing customer expectations
- Continuously improving the process to keep ahead of the competition
- Providing a nonstandard activity to meet a special customer need

Adaptable processes have the capacity to adjust, not only to meet the average customer expectation but to design intelligence into the processes so that they will be able to accommodate individual special needs and expectations. These adaptable processes are not so rigid that they cannot be changed to meet a special customer need without upsetting the process. Remember, the traditional process is designed for the average customer and would therefore satisfy a majority of the customers most of the time. But the individual needs of some customers, as well as changes in the needs of all customers, can be addressed only through an adaptable process.

Imagine, for example, that you have booked an airline seat at your local airport. The formal procedure says that the airline will hold your reservation until 15 minutes before departure time. Unfortunately, the weather is bad, roads are slippery, and you arrive only 5 minutes before takeoff. An adaptable process will *understand* weather conditions and hold your reservation past the 15-minute rule. It would exceed your expectations in order to increase your satisfaction as a customer.

Nordstrom department stores are a good example of how a company can make good use of adaptable procedures. A typical example would be the time a customer came in to pick up his altered suit, and it was not ready as promised. The customer was leaving for Europe and planned to take the suit with him. The salesperson apologized and told the customer that the suit would be shipped overnight by express mail the next day to his hotel in Europe. Not only did the suit arrive on time, but also included in the box were a new shirt and tie, courtesy of Nordstrom.

Adaptable processes are ones designed so that they can be easily changed to meet future customer expectations, to make them more customer friendly, and to reduce processing costs. The one thing that we can be sure of is that the process we are refining today will be changed to make it even better in the future. As the PIT works to improve the process, adaptability and changeability are key considerations.

Of our three key process requirements—effectiveness, efficiency, and adaptability—adaptability is by far the most difficult to measure, but one of the first ones that your customers will complain about. There are a number of good ways to measure process adaptability. For example:

1. The average time it takes to get a special customer request processed compared to standard procedures
2. The percentage of special requests that are turned down

3. The percentage of time special requests are escalated. (In the service industries, the more people a customer has to talk to in order to get a need satisfied, the less chance there is that the person will be satisfied.)

Adaptability requirements should be established at the beginning of the business process cycle so that the improvement activities can consider these parameters and data systems can be established to measure improvements.

WHY MEASURE?

Measurements are key. If you cannot measure it, you cannot control it. If you cannot control it, you cannot manage it. If you cannot manage it, you cannot improve it. It is as simple as that.

Measurements are the starting point for improvements because they enable you to understand what the goals are. Without them, needed change and improvement in the process are severely hindered. You need to develop critical effectiveness, efficiency, and adaptability measurements and targets for the total process. The PIT will use this information to:

- Assess current performance of the process
- Set goals for improvement
- Understand what is important

It is essential to have relevant, specific, measurable, agreed-to, and documented measurements. Now go back and compare these measurements to the process mission statement. Do the measurements support the mission? If not, we have more work to do.

Let's return to the example of the order entry process and summarize the possible measurements for the process:

- Effectiveness
  - Percentage of orders sent within 8 hours
  - Percentage of orders rejected (for incomplete information)
  - Percentage of daily order reports completed on time
- Efficiency
  - Dollar cost per order entered
  - Percentage of time spent on rework
  - Maximum time from order receipt to order entered
- Adaptability
  - Percentage of special orders entered within 8 hours
  - Percentage of special orders processed
  - Percentage of special orders processed at the employee level

PIT GOALS AND TIMETABLE

Up to this point in the process, direction has been provided by the EIT and the process owner. This direction may or may not meet the needs of the people closest to the process. Each member of the PIT must have buy-in to and ownership of the improvement goals and plan.

After the team has reviewed the objectives set by the EIT, it should develop a set of initial goals. These goals should define specific levels of improvement that will be accomplished by specific dates (for example, the cycle time will be reduced to 30 days by May 1, 19—). Setting specific goals requires that a project plan be developed to ensure that the committed timetable is realistic. The PIT should document a project plan that will define the tasks required to complete the assignment and the dates when each task is scheduled for completion. It will be necessary to update this plan as additional data become available.

The project plan is really the process of bringing about improvement. As a result, teams frequently develop flow diagrams that help them define how the process will be analyzed and improved. (For detailed information on flow diagraming, read Chapter 4.) As you lay out your project plan and/or project plan flow diagram, look at and work around key events, their sequence in the process, their dependencies, and how you will know an activity is complete. Detailed outlines should be prepared for the tasks scheduled for completion in the next 90 days. Every 30 days, the detailed 90-day plan should be revised, adding 30 more days. This detailed 90-day plan provides management with an understanding of the resources required to support the improvement process, the people responsible for the activities, and the key target dates. We recommend that Gantt and Milestone charts be used in most cases. For more complex processes with a number of sub-PITs, we find that a PERT chart is a very effective tool.

The initial goals and timetable should be reviewed with the EIT, as should any major revisions.

REGISTER THE PIT

Registering PITs provides systematic control over the entire process. The EIT is responsible for registering all newly formed PIT teams. This registration signifies that upper management sanctions and backs the team's efforts. Registering the PIT also ensures that the business process improvements do not get out of hand. For a PIT to be registered by the EIT, it must meet the following requirements:

- Team membership must be established.
- The team mission statement must be complete.

- The team name must be selected.
- Process boundaries must be defined.
- Effectiveness and efficiency measurements and targets must be defined and agreed to.
  - The PIT project plan must be complete.
  - The 90-day plan must be complete.

When the requirements for registration have been met, the process owner should ask the BPI champion to set up a registration meeting for the PIT with the EIT. At this meeting, the process owner should:

- Introduce the team members
- Discuss the PIT mission statement
- Explain how the PIT defined the boundaries of the process
- Review process measurements and goals
- Review any known problems within the process
- Review the plan and schedule

The EIT evaluates the proposal and also makes sure that the scope of the process is not in conflict with other processes being improved. It is possible that the EIT will recommend modifications to the mission or scope of the PIT. Once the EIT agrees with the proposal, the PIT should receive official approval to proceed. At that point, the EIT should register the team as an official BPI team, responsible for making improvements and reporting on team progress.

SUMMARY

One of the biggest mistakes American organizations make is that they do not take the time required to develop a comprehensive change plan and to get the buy-in from the people who will be affected by the change. In this chapter, we provided you with one way to set the stage for improvement. It takes time to do the job right, but the proper, up-front preparation will pay big dividends when it comes to implementation. Our experience indicates that a good plan, supported by the people who it will affect, minimizes the total cycle time.

Figure 3.8 depicts the cycle time to implement the same change, using two different planning methods, which are:

Method 1. This method typifies the way most American organizations implement change. In this case, little planning precedes the start of implementation. Usually, a small group agrees on the change that should be made, charging ahead and implementing it. As a result, the change meets with a lot of resistance, as the "not-invented here" syndrome rears its ugly head. Not knowing how the

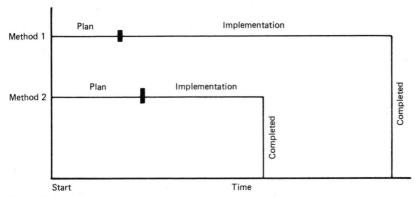

Figure 3.8   Impact of planning on implementing a change.

change is going to impact other organizations, and even their own organization, causes a number of false starts. The people who have to make the change work spend more time explaining why it won't work than they spend making it work. There is a general attitude of "I don't care if it works or not."

Method 2. With this method, much more time is spent developing implementation plans, reviewing them with the people they affect, updating the plans to reflect the input they receive, and letting everyone know what is going to happen. Finally, when the process moves into the implementation stage, all known roadblocks have been considered, and everyone has had a chance to have an impact on the plan; therefore, the general attitude is supportive. The people who are affected feel a commitment to make the change work. There are few restarts, and the implementation time is much shorter. The end result is that method 2 decreases the overall change cycle time and provides a process with a much higher probability of success.

In this chapter, the PIT was formed. It established process boundaries and measurements and set improvement goals for the process. Take your time during this phase of the BPI cycle, and do the job extremely well. It will make the rest of your job much easier.

Be sure that you remember and apply the five P's: **P**roper **P**lanning **P**revents **P**oor **P**erformance.

ADDITIONAL READING

The Systematic Participative Management Series, Harrington, Hurd & Rieker, San Jose, CA, 1983.

4

Flowcharting: Drawing a Process Picture

INTRODUCTION

"One picture is worth a thousand words." If we may modify this age-old proverb and expand it a little to cover your business processes, it might read, "A flowchart is worth a thousand procedures." Flowcharting, also known as *logic* or *flow diagraming*, is an invaluable tool for understanding the inner workings of, and relationships between, business processes. This chapter is designed to help the process improvement team (PIT) member accomplish the twofold task of:

1. Understanding some of the available flowcharting techniques
2. Applying these techniques to understand business processes

Flowcharting is defined as a method of graphically describing an existing process or a proposed new process by using simple symbols, lines, and words to display pictorially the activities and sequence in the process.

WHAT ARE FLOWCHARTS?

Flowcharts graphically represent the activities that make up a process in much the same way that a map represents a particular area. Some advantages of using flowcharts are similar to those of using maps. For example, both flowcharts and maps illustrate how the different elements fit together.

Consider Figure 4.1, a flowchart of the process for hiring a new employee in the fictitious HJH Company. The process begins with a recognition of the need to hire someone and ends with the employee reporting to work. This brief overview of the major activities in the process enables those who understand how to read this story to quickly compare the ways in which HJH's hiring process resembles and differs from that of other companies. For example, you easily can see that HJH emphasizes hiring from inside the organization.

Another advantage is that constructing flowcharts disciplines our thinking. Comparing a flowchart to the actual process activities will highlight the areas in which rules or policies are unclear or are even being violated. Differences between the way an activity is supposed to be conducted and the way it is actually conducted will emerge. Then, with just a few short steps, you and your colleagues will be able to determine how to improve the activity. Flowcharts are a key element in business process improvement (BPI). Good flowcharts highlight the areas in which fuzzy procedures disrupt quality and productivity. Then, because of their ability to clarify complex processes, flowcharts facilitate communication about these problem areas.

FLOWCHARTING OVERVIEW

Flowcharting an entire process, down to the task level, is the basis for analyzing and improving the process. Assigning portions of the process to specific team members will speed up what can be a time-consuming task.

Every situation and/or process will present unique charting problems. The team will have to deal with them as they arise. For instance, existing documentation seldom is sufficient to allow flowcharting of every task and activity without talking to the people performing the tasks. Be careful to distinguish between what the documentation says should be done and what actually is done.

There are many different types of flowcharts, each with its own use. You must understand at least four of these techniques to be effective in the PIT. They are:

1. Block diagrams, which provide a quick overview of a process

2. The American National Standards Institute (ANSI) standard flow-charts, which analyze the detailed interrelationships of a process
    3. Functional flowcharts, which depict the process flow between or-ganizations or areas
    4. Geographic flowcharts, which illustrate the process flow between locations

BLOCK DIAGRAMS

A block diagram, also known as a *block flow diagram*, is the simplest and most prevalent type of flowchart. It provides a quick, uncomplicated view of the process. Figure 4.1 is a block flow diagram that provides an overview of the hiring process. Rectangles and lines with arrows are the major symbols in a block flow diagram. The rectangles represent activities, and the lines with arrows connect the rectangles to show the direction of information flow and/or the relationships among the activities. Some block flow diagrams also include elongated circle start and stop symbols to indicate where the flowchart begins and where it ends.
    Use block diagrams to simplify large, complex processes or to document individual tasks. Include a short phrase within each rectangle describing the activity being performed. Keep these descriptive phrases (activity names) short.
    Let's decode the story told in Figure 4.1.

Activity 1.   A manager recognizes a need for another employee because of high overtime, an employee leaving, etc. To fill this need, he or she must complete the required forms and get the proper approvals.

Activity 2.   The appropriate people review the request for a new employee and approve or reject it. This approval may result in a budget increase. After the necessary approvals are obtained, the approved request is sent to personnel.

Activity 3.   Personnel looks for internal candidates who have been recommended for promotion or transfer who also meet the needs of the job. The HJH Company does not post jobs. A list of candidates, along with their personnel files, is sent to the requesting manager.

Activity 4.   The manager reviews the files and arranges to interview suitable candidates. Then he or she notifies personnel of the results of the review and the interviews.

Activity 5.   If one of the candidates is acceptable, go to activity 10. If not, continue to activity 6.

Activity 6.   Personnel conducts an outside search for candidates by

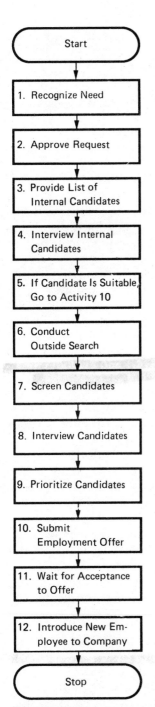

Figure 4.1 Hiring process at HJH Company.

running ads in newspapers, reviewing on-file applications, hiring a search firm, etc.

Activity 7. Personnel reviews potential candidates' applications and conducts screening interviews with the best candidates. Then interviews are set up between the manager and the most promising candidates.

Activity 8. The manager interviews the candidates.

Activity 9. The manager prioritizes the acceptable candidates and sends this list to personnel.

Activity 10. Personnel submits an employment offer to the best candidate.

Activity 11. The company waits for the candidate's response. If the offer is rejected, activities 10 and 11 are repeated for the next candidates on the priority list. Once the offer is accepted, go to activity 12.

Activity 12. Personnel arranges for the employee to report to work, familiarizes him or her with company procedures, and presents the employee to the manager.

As you can see, many activities are performed within each rectangle. If desired, each rectangle can be expanded into a block diagram of its own. Figure 4.2 takes the first activity in Figure 4.1 and explodes it into a more detailed block diagram comprising the following activities:

Activity 1. The manager analyzes the amount of overtime to determine whether a new employee could reduce it sufficiently to offset the cost of his or her salary and benefits.

Activity 2. He or she reviews the procedure for acquiring a new employee.

Activity 3. The manager asks personnel to send blank personnel requisition forms and budget variation forms.

Activity 4. He or she fills out forms.

Activity 5. He or she prepares a job description for the new job.

Activity 6. He or she reviews with the second-level manager and gets a sign-off.

Activity 7. The manager mails the job description, budget change request, and employee requisition form to the controller for approval.

Even in Figure 4.2, some of the activities could be broken down into individual task flowcharts. For example, how to write a job description easily could be a separate block diagram.

Notice that the label description of each activity begins with a verb. Although not mandatory, following this practice is a good general rule. Standard phrasing speeds understanding for the reader. In addition, all

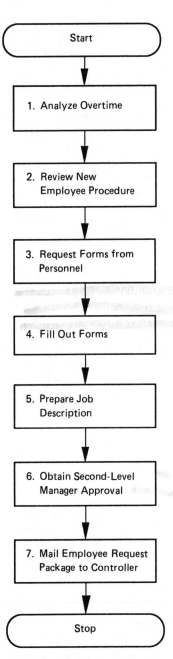

Figure 4.2 Management action required to obtain a new employee approval.

business activities can be described by a verb. Thus, by starting each block label with a verb, you ensure that the label does, in fact, describe a true business activity.

If there are conditional statements in your flowchart, you may not be able to begin every label with a verb. For instance, in Figure 4.1, activity 5 begins with a conditional statement, "If candidate is suitable, go to activity 10." The rule of using a starting verb is still followed—immediately after the conditional statement.

Block diagrams can flow horizontally or vertically. Figures 4.1 and 4.2 flow vertically. Figure 4.3 is a block diagram of a barbecue that is plotted horizontally to the page. Despite the change, the diagram still leads you through the process in a logical way.

Block diagrams provide a quick overview of a process, not a detailed analysis. Normally, they are prepared first to document the magnitude of the process; then another type of flowchart is used to analyze the process in detail.

Typically, many activities and inputs are intentionally not detailed in a block diagram; therefore, a very simple picture of the total process can be drawn. Consider activity 4 in Figure 4.3: "Develop menu." Many activities and inputs must go into developing a menu for the barbecue. The typical inputs required are:

1. The amount of money to be spent
2. Guests' preferences
3. What we prepare well

The typical activities include:

1. Listing the items to be served
2. Listing the materials needed for the menu
3. Getting money to pay for the food and condiments

It is easy to see how each of the blocks in the block diagram can be exploded to provide a detailed picture of how the activity is performed. Don't worry if all the process details are not documented in the block diagram. The detailed activities will come later in the flowchart process.

Figure 3.3 in Chapter 3 is another way of flowcharting the summer barbecue process depicted in Figure 4.3. Figure 3.3 is plotted vertically, and responsibility for each activity has been assigned to a specific person or persons. The name or title of the person responsible for the activity is indicated in the open-ended rectangle. This symbol is called an *annotation symbol* since it is used to provide additional information about the activity. A broken line leads away from the activity to the annotation symbol. The broken line is used so that the reader will not mistake it for a direction flow line. The arrow leads away from the block diagram activity and points to the person or persons responsible for that activity. When your organiza-

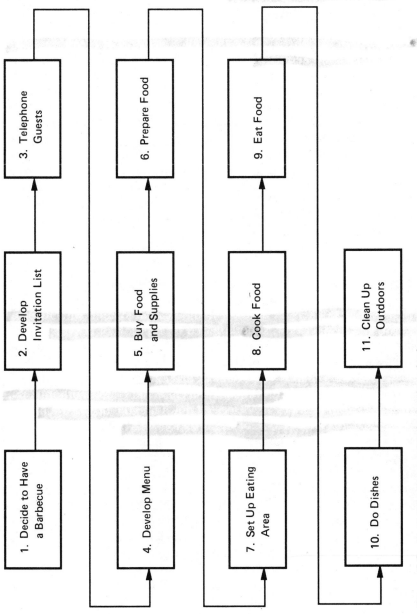

Figure 4.3 Block flow diagram for conducting a barbecue.

1. Decide to Have a Barbecue

2. Develop Invitation List

3. Telephone Guests

4. Develop Menu

5. Buy Food and Supplies

6. Prepare Food

7. Set Up Eating Area

8. Cook Food

9. Eat Food

10. Do Dishes

11. Clean Up Outdoors

tion uses block flow diagrams to chart a set of business activities, you may indicate responsibilities differently. You may use the name of a department, the job titles of employees and managers, or the actual names of the individuals in denoting responsibilities.

The purpose of flowcharting is to paint a picture that is easy for your team to understand and use. You can modify rules, such as starting each activity name with a verb or using annotation symbols in place of the activity owner's name within the activity rectangle, if doing so significantly improves the understandability and use of the flowchart. However, given that any nonstandard deviation may confuse other people within the organization using the flowchart at a later date, it is a good idea to have the executive improvement team (EIT) establish a complete list of symbols at the beginning of the BPI activity to minimize deviations.

It is good practice to start your business process flowcharting by block diagraming the process. The block diagram can be used to help define which of the other flowcharts best provides a detailed understanding of the tasks within the process.

BLOCK DIAGRAMING ACTIVITIES AND INFORMATION

A process is also likely to have a communication system, with its own separate and distinct flow, superimposed on the flow of activities. This communication system also must be recognized, flowcharted, and understood as an integral part of the process operations.

An organization chart is a type of block diagram. In this case, the reporting structure is pictured. An organization chart shows how authority, responsibilities, and activities are delegated down into the organization.

Figure 4.4 presents a typical organization chart. The organization flow is represented by solid lines, while the communication system is indicated by dotted lines. The communication flow in most organizations is an essential, but complex, part of the organizational structure. A good communication system flows up, down, and sideways. Frequently, a communication flow line will have arrows on both ends, signifying two-way communication. A typical two-way communication would be a meeting at which everyone is invited to contribute to the discussion. As you can see, the communication flow is much more complex than the organization flow.

Figure 4.4 reveals some interesting patterns. Mid-level department A2 is not part of upper management's communication flow and, as a result, holds meetings with departments A1 and A3 in an attempt to communicate its concerns to upper management. Typically, A2 meets first with A1 to obtain a status report, and then meets with A3 to verify the information obtained at the first meeting.

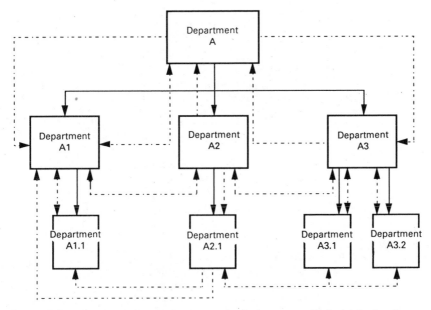

Figure 4.4   A block diagram with its communication systems added with broken lines.

Unfortunately, this pattern is repeated at lower levels because, while the manager of department A2 communicates the verified data to the first-line department A2.1 that reports to him or her, he or she never solicits input from A2.1, developing another communication void. As a result, the manager of department A2.1 has developed a very active communication system with the other first-line departments in the hope that the department's activities and concerns reach upper and middle management.

STANDARD FLOWCHART SYMBOLS

Before examining the remaining three types of flowcharts, we should define some additional symbols. The most effective flowcharts use only widely known, standard symbols. Think about how much easier it is to read a road map when you are familiar with the meaning of each symbol and what a nuisance it is to have some strange, unfamiliar shape in the area of the map you are using to make a decision about your travel plans.

The flowchart is one of the oldest of all the design aids available. For simplicity, we will review only 12 of the most common symbols, most of which are published by ANSI (see Figure 4.5).

Symbol	Meaning
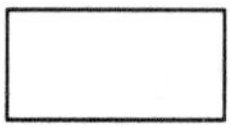	*Operation: Rectangle.* Use this symbol whenever a change in an item occurs. The change may result from the expenditure of labor, a machine activity, or a combination of both. It is used to denote activity of any kind, from drilling a hole to computer data processing. It is the correct symbol to use when no other one is appropriate. Normally, you should include a short description of the activity in the rectangle.
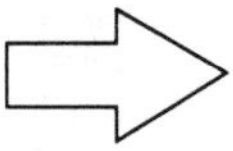	*Movement/transportation: Fat arrow.* Use a fat arrow to indicate movement of the output between locations (e.g., sending parts to stock, mailing a letter).
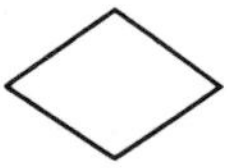	*Decision point: Diamond.* Put a diamond at the point in the process at which a decision must be made. The next series of activities will vary based on this decision. For example, "If the letter is correct, it will be signed. If it is incorrect, it will be retyped." Typically, the outputs from the diamond are marked with the options (e.g., YES-NO, TRUE-FALSE).
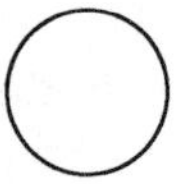	*Inspection: Big circle.* Use a big circle to signify that the process flow has stopped so that the quality of the output can be evaluated. It typically involves an inspection conducted by someone other than the person who performed the previous activity. It also can represent the point at which an approval signature is required.
	*Paper documents: Wiggle-bottomed rectangle.* Use this symbol to show when the output from an activity included information recorded on paper (e.g., written reports, letters, or computer printouts).
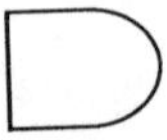	*Delay: Blunted rectangle.* Use this symbol, sometimes called a bullet, when an item or person must wait, or when an item is placed in temporary storage before the next scheduled activity is performed (e.g., waiting for an airplane, waiting for a signature).

Figure 4.5   Standard flowchart symbols.

Symbol	Meaning
	Storage: Triangle. Use a triangle when a controlled storage condition exists and an order or requisition is required to remove the item for the next scheduled activity. This symbol is used most often to show that output is in storage waiting for a customer. The object of a continuous-flow process is to eliminate all the triangles and blunt rectangles from the process flowchart. In a business process, the triangle would be used to show the status of a purchase requisition being held by purchasing, waiting for finance to verify that the item was in the approved budget.
	Annotation: Open rectangle. Use an open rectangle connected to the flowchart by a dotted line to record additional information about the symbol to which it is connected. For example, in a complex flowchart plotted on many sheets of paper, this symbol could be connected to a small circle to provide the page number where the inputs will reenter the process. Another way to use an open rectangle is to identify who is responsible for performing an activity or the document that controls the activity. The open rectangle is connected to the flowchart with a dotted line so that it will not be confused with a line arrow that denotes activity flow.
	Direction of flow: Arrow. Use an arrow to denote the direction and order of process steps. An arrow is used for movement from one symbol to another. The arrow denotes direction—up, down, or sideways. ANSI indicates that the arrowhead is not necessary when the direction flow is from top to bottom or from left to right. However, to avoid misinterpretation by others who may not be as familiar with flowchart symbols, it is recommended that you always use arrowheads.

Figure 4.5   (Continued)

Symbol	Meaning
	*Transmission: Interrupted arrow.* Use an interrupted arrow to identify when immediate transmission of information occurs (e.g., electronic data transfer, fax, telephone call).
	Connector: Small circle. Use a small circle with a letter inside it at the end of a flowchart to indicate that the output from that part of the flowchart will serve as the input to another flowchart. This symbol often is used when there is not enough room to get the entire flowchart on one piece of paper. An arrowhead pointing at the small circle denotes that the circle is an output. An arrowhead facing away from the small circle denotes that it is an input. Each different output should have a different letter designation. Any output can reenter the process at a number of different points.
	*Boundaries: Elongated circle.* Use an elongated circle to show the beginning and end of the process. Normally, the word *start* or *beginning/stop* or *end* is included within the symbol.

Figure 4.5 (Continued)

The 12 symbols listed in Figure 4.5 are not meant to be a complete list of flowchart symbols, but they are the minimum you will need to adequately flowchart your business process. As you learn more about flowcharting, you can expand the number of symbols you use to cover your specific field and needs.

ANSI STANDARD FLOWCHART

An ANSI standard flowchart provides a detailed understanding of a process that greatly exceeds that of a block diagram. In fact, a block diagram often is the starting point, and a standard flowchart is used to expand the activities within each block to the desired level of detail. Each task in the process under study can be detailed to the point that the standard flowchart can be used as part of the training manual for a new employee. For most BPI activities, this type of detail is done on an exception-only basis during the improvement phase. Detailed flowcharting is done only when the process nears world-class quality, to ensure that the improvements are not lost over time.

People follow many different processes throughout their daily lives. As an example, a person takes on a particular routine for such simple tasks as eating breakfast, taking a shower, or enjoying a Saturday morning. Most of these processes are not even thought about. Some processes involve other people to such a degree that we don't think about our own involvement. One such process might be that of getting a haircut from the friendly corner barber and/or going fishing. This process is flowcharted in Figures 4.6 and 4.7.

The standard flowchart in Figure 4.6 shows diamonds as decision symbols representing points at which different paths may be taken. Notice that the words *yes* and *no* are used to clarify alternatives. The small circles are connector symbols leading you to the second page of the chart (Figure 4.7).

A SIMPLE BUSINESS PROCESS FLOWCHART

While the flowcharts in Figures 4.6 and 4.7 are very simple, charting a business process requires careful attention. Consider the manager of a large retail store in a big city. The procedures he or she must follow can become quite complicated. He or she may have a large staff, delegate authority, supervise various departments, and so on. Each supervisor has sales reports to complete and check against inventory changes. The manager must provide each supervisor with instructions to help ease the work load and promote uniformity among the different departments. This, in turn, helps the accounting department.

A typical procedure for a supervisor might include:

1. Choosing the weekly sales total for an employee; reading the value of price items from column X and the value of sales items from column Y

2. Figuring out the X commission by multiplying the value in column X by 10 percent

3. Figuring out the Y commission by multiplying the value in column Y by 5 percent

4. Computing the total due: $50.00 + X commission + Y commission

5. Entering the total pay opposite the employee's name in the payroll ledger

6. Returning to activity 1, and repeating this for the other employees

Figure 4.8 flowcharts the procedure for calculating employees' weekly commissions. The activities in the procedure are listed beside each symbol in the flowchart to help people understand the details of the flowchart. Unfortunately, this is not usually practical on complex flowcharts.

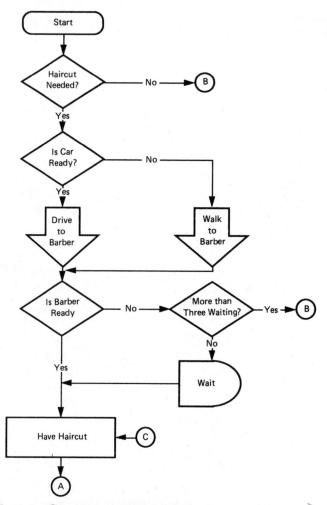

Figure 4.6 A standard flowchart of the first part of the process of getting a haircut and/or going fishing.

The first five activities in Figure 4.8 follow activities 1 through 5 of the written procedure on page 99. Notice, however, that the flowchart allows for an activity not accounted for in the written procedure (i.e., eventually, the weekly sales totals for all employees will have been processed, and the procedure need not be repeated). Flowcharting the process, in this case, helps us to discover that activity 6 should be rewritten as follows:

6. If the weekly sales totals for more employees must be calculated, go to activity 1. Otherwise, stop.

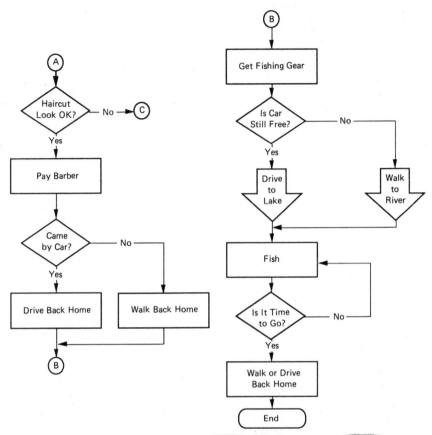

Figure 4.7 Two standard flowcharts of different parts of the process of getting a haircut and/or going fishing.

This simple flowchart clearly and accurately depicts the activities involved in the procedure and the sequence in which they are to be carried out.

FUNCTIONAL FLOWCHART

A functional flowchart is another type of flowchart. It pictures the movement between different work units, an additional dimension that is particularly valuable when total cycle time is a problem. A functional flowchart uses either block or standard flowchart symbols.

A functional flowchart identifies how vertically oriented functional departments affect a process flowing horizontally across an organization. If a process always was contained within a single department

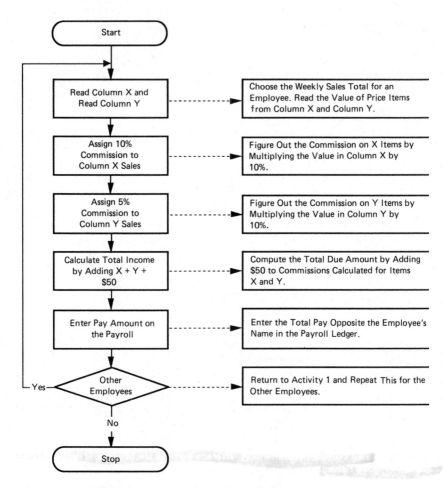

Figure 4.8 Paying commission flowchart and procedures.

and didn't cross over to other territories, a manager's life would be much easier. However, in most companies, the functional or vertical organization is a way of life, because it provides a highly trained competency center that cannot be equaled using a process or product organization.

Figure 4.9 is a standard functional flowchart of the hiring process that was block diagramed in Figure 4.1 (activities 1 through 5). To keep the flowchart simple, we have used only three of the standard symbols. We also have expanded the first 5 activities in Figure 4.1 to 15 activities and separated them by the area performing them. The 15 activities are listed on pages 103 and 104.

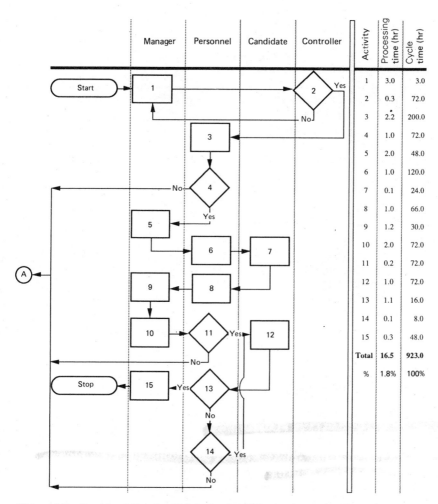

	Activity	Processing time (hr)	Cycle time (hr)
	1	3.0	3.0
	2	0.3	72.0
	3	2.2	200.0
	4	1.0	72.0
	5	2.0	48.0
	6	1.0	120.0
	7	0.1	24.0
	8	1.0	66.0
	9	1.2	30.0
	10	2.0	72.0
	11	0.2	72.0
	12	1.0	72.0
	13	1.1	16.0
	14	0.1	8.0
	15	0.3	48.0
	Total	16.5	923.0
	%	1.8%	100%

Figure 4.9 Functional flowchart of the internal job search process.

	Activity	Responsible area
1.	Recognize need. Complete payback analysis. Prepare personnel requisition. Prepare budget request.	Manager
2.	Evaluate budget. If yes, sign personnel requisition slip. If no, return total package with reject letter to manager.	Controller
3.	Conduct in-house search.	Personnel
4.	If in-house candidates exist, provide list to management. If not, start outside hiring procedure.	Personnel

	Activity	Responsible area
5.	Review candidates' paperwork and prepare a list of candidates to be interviewed.	Manager
6.	Have candidates' managers review job with the employees and determine which employees are interested in the position.	Personnel
7.	Notify personnel of candidates interested in being interviewed.	Candidates
8.	Set up meeting between manager and candidates.	Personnel
9.	Interview candidates and review details of job.	Manager
10.	Notify personnel of interview results.	Manager
11.	If acceptable candidate is available, make job offer. If not, start outside hiring process.	Personnel
12.	Evaluate job offer and notify personnel of candidate's decision.	Candidate
13.	If yes, notify manager that the job has been filled. If no, go to activity 14.	Personnel
14.	Were there other acceptable candidates? If yes, go to activity 12. If no, start outside hiring process.	Personnel
15.	Have new manager contact candidate's present manager and arrange for the candidate to report to work.	Manager

FUNCTIONAL TIME-LINE FLOWCHART

A functional time-line flowchart adds processing and cycle time to the standard functional flowchart. This flowchart offers some valuable insights when you are doing a poor-quality cost analysis to determine how much money the organization is losing because the process is not efficient and effective. Adding a time value to the already-defined functions interacting within the process makes it easy to identify areas of waste and delay.

Time is monitored in two ways. First, the time required to perform the activity is recorded in the column entitled "Processing time (hr)." The column beside it is the cycle time (i.e., the time between when the last activity was completed and the time this activity is completed). Usually, there is a major difference between the sum of the individual processing hours and the cycle time for the total process. This difference is due to waiting and transportation time.

In Figure 4.9, while the total processing time is only 16.5 hours, the total cycle time is 923.0 hours. Performing all the activities required only 1.8 percent of the total time that it took to fill one job. The cycle time analysis shows why it takes so much time to get even the simplest job done.

One common error is to focus on reducing processing time and to ignore cycle time. The result is focusing our activities on reducing costs, without considering the business from our customers' viewpoints. Customers do not see processing time; they see only cycle time (response time). To meet our needs, we work on reducing processing time. To have happy customers, we must reduce cycle time.

In one sales process, IBM was able to reduce processing time by 30 percent, thereby reducing costs by 25 percent. At the same time, the company reduced cycle time by 75 percent. An unplanned-for side effect was a more than 300 percent increase in sales (65 percent sales closure). There is no doubt that there is a direct correlation between cycle time, customer satisfaction, and increased profits.

The time-line flow concept can be applied to all types of flowcharts. Often, elapsed time is recorded using the time that has elapsed from the time the first activity in the process started. If this method were used in Figure 4.9, the elapsed time recorded adjacent to activity 3 would be the sum of the time recorded for activities 1, 2, and 3, or 275 hours.

GEOGRAPHIC FLOWCHART

A geographic or physical layout flowchart analyzes the physical flow of activities. It helps to minimize the time wasted while work output and/or resources are moved between activities. Figure 4.10 presents a geographic flowchart of how a new employee spends his or her first day at the HJH Company. It starts with a geographic layout of the buildings at HJH headquarters. Laid over the geographic layout, using broken lines, is the movement of the new employee on his or her first day. To help you understand how to use this chart, we will follow the new employee through the first day:

1. New employee signs in at lobby and asks receptionist to call personnel.

2. Personnel placement representative greets new employee and takes him or her to personnel department to review pertinent procedures.

3. Placement representative takes new employee to medical department to fill out medical forms and make appointment with nurse for required tests.

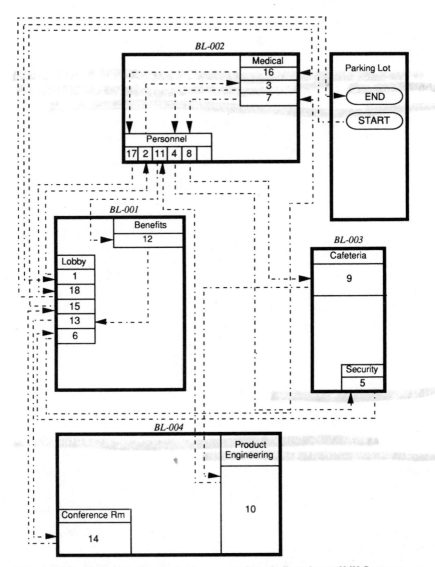

Figure 4.10   Geographic flowchart of a new employee's first day at HJH Company.

4. New employee returns to personnel department to fill out payroll forms.

5. New employee and the placement representative go to security for pictures and temporary identification badge.

6. New employee returns to lobby to wait for appointment with nurse. He or she can go unescorted now that he or she has a temporary badge.

7. New employee goes to medical for blood tests and makes appointment for physical exam with doctor.

8. New employee returns to personnel department per instructions.

9. Placement representative takes new employee to lunch.

10. Placement representative takes new employee to meet his or her new manager and tour the department.

11. New employee goes to personnel so that personnel can take him or her to benefits.

12. New employee reviews benefits package and selects benefit plan.

13. New employee goes to lobby to wait for the new employee orientation meeting.

14. New employee attends new employee orientation meeting.

15. New employee returns to lobby to wait for appointment with doctor.

16. New employee goes to medical for appointment with doctor and returns to personnel.

17. Personnel reviews new employee checklist and calls medical to find out whether exam results are favorable.

18. New employee returns to lobby, turns in temporary badge, and signs out.

First impressions are key. How do you think new employees feel about this company after a day of "hurry up and wait"? Probably, they are questioning whether they made the right decisions in joining the company. Analyzing this flowchart quickly reveals wasted motion and time. For example, if new employees reported to a special waiting room in personnel, the amount of time the employees and the personnel placement representative expended during the day would be greatly reduced. If personnel gave out the temporary badges, personnel would not have to escort the employees to other departments.

Let's think about what can be done to refine the flow and make better use of the new employees' and the personnel placement representative's time.

1. Should the physical be conducted before the employee reports to work? Isn't it part of the search process, not the indoctrination process? If the new employee had left another job to join your firm and then failed the physical, what is your company's obligation to that person? How much would it slow down the process to get a new employee on board? Obviously, the physical should have been conducted before the new employee's first day.

2. Have the employee report first to a small waiting room in personnel. At that time, the personnel placement representative can provide a temporary badge.

3. From personnel, the new employee should go directly to benefits.

4. The employee indoctrination meeting should be held right after the meeting with the benefits department.

5. The new employee's manager should meet him or her at the end of the indoctrination meeting and proceed to security. A picture of the new employee should be taken for the permanent security badge.

6. The manager then should take the new employee to lunch.

7. The manager should escort the new employee to the work area and proceed with job training.

8. The new employee should keep the temporary badge until a permanent badge is available. This allows the employee to go home directly from work. When the employee's permanent badge is available, the manager should mail the temporary badge back to personnel.

Figure 4.11 shows the new geographic flowchart. In addition to simplifying the work flow, the new employee is now in his or her work area for the second half of the day. The result is a more efficient process that leaves the new employee with a much better impression of the company.

Geographic flowcharting is a useful tool for evaluating department layout and paperwork flow, and for analyzing product flow, by identifying excessive travel and storage delays. In business processes, geographic flowcharting helps in analyzing traffic patterns around busy areas like file cabinets, computers, and copiers.

TAKING AN INFORMATION-PROCESSING VIEW

In addition to the four basic flowcharts we have covered already (block diagrams, ANSI standard flowcharts, functional flowcharts, and geographic flowcharts), there are information diagrams, often with their own set of symbols. As a rule, these are of more interest to computer programmers and automated systems analysts than to managers and employees charting business activities. The two books listed at the end of this chapter discuss some of these tools.

You can consider these types of flowcharts as diagrams that follow information through a process. As you prepare flowcharts, think of your organizational activities in terms of information processing. Begin with your organization's files. They are valuable because they contain information that is changed or used by your business processes.

Next, consider your employees. You and your coworkers have skills of various levels and types. Obviously, even a single worker's knowledge is substantially more sophisticated than the information in a file. But the principle still holds: An employee's value to an organization depends on

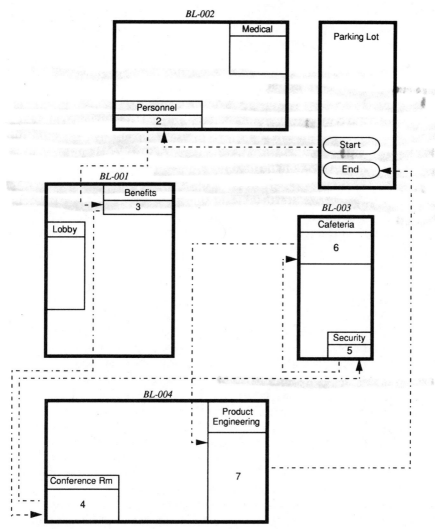

Figure 4.11 Revised geographic flowchart of a new employee's first day at HJH Company.

his or her contributions of information. Whether it's how to load a pallet, introduce a new product, or resolve a conflict, information is a resource. This is particularly true in the service industries which, in 1989, employed more than 70 percent of workers in the United States. All of them can be considered information processors and providers.

Taking an information-processing view when preparing your flowcharts will create a common focus on getting and using quality input in order to produce quality output. At the same time, an information-processing view helps people decide how to draw flowcharts and which elements to include. More specifically, you should:

1. Feature the parts of the process for which information validity and reliability are most important.

2. Consider the three different information-processing dimensions of business processes: what information is processed, what activities are involved in processing the information, and which elements control other elements. If your flowchart doesn't have the impact you want, try drawing it to feature one of the other dimensions.

3. Remember that organizations consist of people, and whenever people are involved, information transmission and processing are complicated. Consequently, it is better to draw several easily understandable flowcharts than one comprehensive, but incomprehensible, master chart.

DATA DICTIONARY

Many of you may never need to use a data dictionary because your flowcharts will be reasonably uncomplicated and straightforward. There will be some of you, though, who will need to go into considerable detail covering a broad range of activities. In this case, the use of a data dictionary becomes necessary in order to be sure that all labels and definitions are clear and understood.

The most effective flowcharts use words and phrases that people will easily understand, and they include only widely known, standard symbols. Often, an accompanying glossary of terms, known as a *data dictionary* by information-processing professionals, helps. Each entry in the dictionary refers to a label used in the flowcharts.

A data dictionary serves a number of reference purposes. For example, it alerts you to database homonyms. A *database* is a collection of information inside an organization's files. (Often these files are computerized.) Homonyms exist when the same label refers to different items. Consider, for example, the label "Enter employee ID." On one flowchart, this might mean, "Record the employee's social security number on a form." On another flowchart it might mean, "Type the employee's name into a computer system and wait for the system to verify the entry." Because of their multiple meanings, database homonyms can cause confusion in a set of flowcharts.

Homonyms occur because flowchart labels must be brief. You don't have space for a detailed explanation on the chart itself, but you can include the definition in the data dictionary. Checking the dictionary before selecting a label will tell you whether there are other ways in which your label is being used already. If this is so, you might select another label or take special measures to ensure that people using your flowchart know what you really mean.

You also can use the data dictionary for assistance with database synonyms—when different labels have identical definitions. For example, *receivables* might refer to the same thing as *sales collectible*.

As with homonyms, database synonyms may be necessary. People prefer to use familiar terms when constructing their flowcharts, and employees in different parts of the organization may have different words for an identical item. Recognizing the value of familiarity, information-processing professionals call database synonyms that are acceptable *aliases*.

Yet synonyms must be identified. Otherwise, a team drawing an overview flowchart with activities from the accounting department and the sales department might include unnecessary duplication. That's where developing a data dictionary can help. When the accounting department is asked to define the receivables file, and the sales department is asked to define the sales collectible file, the team creating the flowchart will discover that the two files are the same.

In addition, a data dictionary can include detailed information, beyond a definition, about the activity represented on the flowchart. In searching for ways to improve business activities, a team may be interested in how many records a certain file contains or how many items are processed each week through a certain activity. The team also may want to know on which flowcharts a given label appears to accurately evaluate the implications of a change. All these details can be stored in the data dictionary.

Data dictionaries can be maintained manually or on a computer system. With a computer system, you can more easily revise, arrange, and locate information. If you create your flowcharts with a computer, automating the data dictionary has even further advantages, since you can develop a system of automated cross-references between the charts and the dictionary.

SUMMARY

Flowcharting is a key tool for understanding business processes. Laying out a process on a piece of paper in an easily understandable format often sets the stage for major process improvement. It is also an effective tool for analyzing the impact of proposed changes. Many business process flowcharts become very complex, often covering an entire wall, but the understanding the PIT gains from this type of analysis is well worth the effort. In the case of new processes, flowcharts should precede the preparation of the procedures.

To improve the quality of their products and services, many businesspeople have used flowcharting techniques with enviable results. Others, however, have been less successful. Generally, this happens because they view their flowcharts as the end of, rather than the means to, what they are seeking. It is an easy mistake to make. Compared to some techniques for improving quality and productivity, flowcharting is easy to understand and use. Furthermore, in their enthusiasm for improvement, some people are tempted to flowchart in detail every process they can find. Fortunately, however, such diligence is rarely necessary.

Flowcharts serve one main purpose: to document a process in order to identify areas in need of improvement. The "magic" doesn't come from documenting the process but from analyzing it—and that is where you should focus most of your efforts. Remember, the purpose of flowcharting and the following analysis is to gain enough knowledge to define and implement process improvements. It should not become an end unto itself.

Flowcharts are tools. It is in the BPI activities following flowcharting that their full value is realized. However, the flowcharting process itself prepares people for the productive changes ahead:

1. Those who participate in creating the flowcharts recognize their own competence and influence. They now know how their contributions serve to empower their coworkers. They are proud that their role is documented on a diagram that others will consult.

2. People see that the value of their performance affects how others use the output. This stimulates curiosity about customers' expectations and strengthens ties between employees and customers.

3. In creating flowcharts, people gain understanding of each other's jobs, resulting in increased cooperation in the work environment. Building flowcharts builds teamwork.

4. As the flowchart grows, participants are inspired by the available sources of assistance and support. The message of the flowchart is that there are power and companionship in the organization.

5. At the same time, individual accountability blooms. The flowchart triggers improvement efforts, adherence to standards of quality, and commitment to reduce process variations.

6. Objective setting is facilitated, even in those parts of the organization that have resisted performance measurement or where people have argued about what are legitimate, realistic objectives.

Throughout all of this, flowcharts focus attention on opportunities for change. As BPIs occur, your team will recognize where the charts are no longer accurate and where revisions are necessary. In addition, you'll create new versions of flowcharts as you and your coworkers become more skilled at constructing them. Some of this is growth in technical and artistic talents. But a much more important part is developing conceptual talents. The people in your organization will begin to view business activities more systematically and more creatively. As you build flowcharts, and check their accuracy, you'll become more sensitive to ways in which you can make your business better.

ADDITIONAL READING

Jeffrey, D. R., and M. J. Lawrence, *Systems Analysis and Design*, Prentice Hall, Englewood Cliffs, NJ, 1984.
Modell, M. E., *A Professional's Guide to Systems Analysis*, McGraw-Hill, New York, 1988.

5

Understanding the Process Characteristics

INTRODUCTION

"Knowledge is the only instrument of production that is not subject to diminishing returns," noted J. M. Clark in the *Journal of Political Economy*. That can be interpreted to mean that the more we understand business processes, the more we can improve them. To do that, we must clearly understand several characteristics of business processes:

- *Flow.* The methods for transforming input into output
- *Effectiveness.* How well customer expectations are met
- *Efficiency.* How well resources are used to produce an output
- *Cycle time.* The time taken for the transformation from input to final output
- *Cost.* The expense of the entire process

Understanding these process characteristics is essential for three reasons. First, it helps identify key problem areas within the process. This information will provide the basis for streamlining the process. Second, it provides the database needed for us to make informed decisions about improvements. We need to see the impact of changes not only on individual activities but also on the process as a whole and on the departments involved. And third, it is the basis for setting improvement targets and for evaluating results.

In Chapter 4, we reviewed how to flowchart a process based on documentation and the process improvement team's (PIT's) understanding of the process. A flowchart is the first step in changing a process. However, process documentation may not always reflect real life because of errors or misunderstandings. Therefore you should verify the accuracy of the process documentation, a topic discussed later in this chapter. You also should understand and gather information about other process characteristics (e.g., quality, cycle time, and cost). This provides valuable information about the location of existing problems.

THE EMPLOYEE AND THE PROCESS

We have talked about the process as a *cold* thing—procedures, equipment, flowcharts, and techniques. The process is brought to life by people. Our people make the process work; without them, we have nothing. We need to understand how the people who bring life to the process feel about the process. What gets in their way? What are the parts of the process that they like? What bores them? The end process has to be a homogeneous marriage of people and methodologies, in which the equipment is the slave to the people, not the other way around.

In the long run, the success of the business process improvement (BPI) activities will depend on the degree to which our people embrace the changes made to the process. Without considering the human side of the process, the PIT cannot be successful. There is only one way to gain the required understanding of the human side of the process, and the talents and the limitations our employees have, and that is to get out into the work environment. Talk to them. Ask for their opinions and ideas. Then, implement their suggestions. If the people are involved, the end results will be much better and far easier to implement.

PROCESS WALK-THROUGH

In Chapter 4, the PIT developed a flowchart of the entire business process. Very often, however, the process documented by the PIT is not what is really happening in the organization. Employees deviate from the process for a number of reasons; for example:

1. They misunderstand the procedures.
2. They do not know about the procedures.

3. They find a better way of doing things.
4. The documented method is too hard to do.
5. They are not trained.
6. They were trained to do the activity in a different way.
7. They do not have the necessary tools.
8. They do not have adequate time.
9. Someone told them to do it differently.
10. They don't understand why they should follow the procedures.

It's also possible that the process flowchart was assembled incorrectly by the PIT. The only way that you can really understand what is happening in the business processes is to personally follow the work flow, discussing and observing what is going on. This is called a *process walk-through*.

To conduct a walk-through, the PIT should physically follow the process as documented in the flowchart from beginning to end(s). The PIT should observe the process at the task level. The team needs to know and understand what is being done and why it is being done. While the PIT performs the walk-through, it will be able to gather additional information about existing problems and roadblocks to change and to make suggestions for improvements.

To prepare for the process walk-through, the PIT should assign team members (usually groups of two or three people) to different parts of the process. Typically, one member of the walk-through team (WTT) will be from the department in which the activity is being performed. When the team is made up of three people, it is also a good idea to have a customer of the department in which the activity is taking place as a member of the WTT. The people who are assigned to the WTT should have some understanding of the activity they will be evaluating. This facilitates review and verification of the process flow. Each WTT should:

- Become very familiar with all relevant, existing process documentation
- Arrange with the department manager to interview his or her people
- Interview a sample of the people performing the task to fully understand what is occurring in the process
- Compare the way different people do the same job to determine what the best standard operation should be

Preparation is the key to a successful walk-through. The WTT really must understand what should be happening in the process and be able to talk in terms that are relevant to the person performing the activity. This requires a lot of work prior to the interview process.

To prepare for conducting the evaluations, the WTT should collect the detailed, documented task descriptions for the activities that are going to be studied. Each team member should become very familiar with these task descriptions. If task descriptions are not available, much more information will need to be recorded during the walk-through.

During the process walk-through, the WTT will have an opportunity to develop a list of the tasks required to support each activity. For example, let's look at the tasks required to support the activity of typing a letter:

1. Read handwritten memo
2. Check punctuation
3. Check spelling and proper names and obtain mailing address
4. Assign file reference number to document
5. Ensure that proper letterhead paper is inserted in printer
6. Turn on word processor, load program, and insert proper disk
7. Type letter
8. Use spell check
9. Proofread letter
10. Print letter
11. Review printed letter to ensure that it is positioned correctly on the paper
12. Place in manager's incoming mail

Doing the task analysis and documentation often reveals new suppliers to the process. It also provides keys to how to improve the process. The task analysis should be prepared in conjunction with the person performing the activity because that is the only way to know how the activity is being performed; the person performing the activity has the best understanding of what is involved.

The PIT should prepare a *process walk-through questionnaire* to collect needed information about the process. Typical questions might be:

- What are the required inputs?
- How were you trained?
- What do you do?
- How do you know your output is good?
- What feedback do you receive?
- Who are your customers?
- What keeps you from doing error-free work?
- What can be done to make your job easier?
- How do you let your suppliers know how well they are performing?
- How is your output used?
- What would happen if you did not do the job?
- Have you reviewed your job description?

- What would happen if each of your suppliers stopped providing you with input?
- What would you change if you were the manager?

A member of the WTT should schedule meetings with the individual employees through the department managers. Care should be taken to ensure that the right types of input are at the work station during the meeting so that the team can observe the activity under real conditions. In some cases, the PIT will set up a pilot run and follow it through the process. This is a good practice but cannot be used to measure cycle time because it will receive very special handling throughout the entire process.

Before the WTT starts the interview, the role of each team member should be defined. One member of the team should be a scribe—usually the team member from the work area makes the best scribe. The other team members should also record notes but usually not in the same detail as the scribe, since their role is to interview the employees. These notes will be very helpful later since the team members from the work area often have preconceived opinions about how the job is being performed. These preconceived opinions sometimes prevent him or her from noting important information that is often obvious to a person not so involved with the process.

Another element of success is the way the interview is conducted. Many employees feel threatened and intimidated being interviewed by the PIT. The small WTTs help, but that is not enough. Dress to fit the environment. A black suit, white shirt, and tie are totally inappropriate for interviews conducted in a warehouse or service center. Take time to put the interviewee at ease. Before you ask questions, explain why you are talking to him or her. Show the interviewee the flowchart and explain how he or she fits into the big picture. Interviewing is an art. You should have interview training to get satisfactory results. The appendix to this book provides interview guidelines. Before the interview cycle begins, all PIT members should study, not just read, this appendix. A 4-hour class on interviewing methods is also very helpful.

Immediately following each interview, the team should schedule a short meeting to review the interview and agree on:

- Task flow
- Required inputs
- Measurements
- Feedback systems
- Conformance to procedure and to other employees
- Major problems
- Cycle-time estimates
- Value-added content
- Training requirements

It is often helpful to flowchart the tasks so that the team will gain a better understanding of the activity being evaluated and be in a better position to report its findings to the PIT.

The outcomes from the process walk-through should include:

- Differences between the documented process and present practice
- Differences between the way employees are performing the activity
- Identification of employees needing retraining
- Suggested improvements to the process (generated by the people performing the process)
- Process measurement points and measurements
- Activities that need to be documented
- Process problems
- Roadblocks to process improvement
- Suppliers that have input into the process
- Internal process requirements
- Elapsed cycle time and activity cycle time
- New training programs required to support the present process
- How suppliers should receive feedback data
- A task flowchart

We find that it is a good practice to review the findings with the interviewees to be sure the team did not misinterpret their comments. A summary of the interviews of all members in a department should be reviewed with the department manager before they are reviewed with the PIT. The department manager and the WTT should agree on what action will be taken to eliminate differences between employees and/or between practice and procedure.

Clearly identify any differences between what is supposed to occur and current practice. Determine why these differences exist. Analyze why everyone is not doing the same job the same way. Standardization is the key to improvement and the first task that must be undertaken. Select a way of performing an activity that provides the best results and use it consistently until you make a major change to the process. It is important for everyone to do the same job in the same way. The output must be predictable before you change the process.

When the walk-through is complete, each WTT should present its findings to the PIT. This provides the total PIT with a better understanding of the process. Based on our experience, we find that the best way to present this data to the PIT is to follow the flowchart, starting at the beginning and working your way through to the end, marking up the flowchart as you go along.

It is important to be able to *readily* identify all activities and tasks not being performed per prescribed procedures. One method is to circle these areas of concern using a yellow highlighter on the flowchart. Action plans should be developed to either change the procedure or bring the activity in line with the procedure. In addition, the walk-through normally will identify a number of problems that need immediate action and others that can be addressed as the process is revamped. Action plans should be established for all the hot problems. These action plans should include:

- The action to be taken
- When the action will be taken
- The individual who will take the action
- How the PIT will know that the action was taken and that it is effective at eliminating the problem

Based on the data collected, the PIT can summarize the key problems. It is important to attempt to segregate quality problems into *occasional* or *chronic* problems. Occasional problems occur only sporadically, tend to stand out, and are quickly corrected. Chronic problems, on the other hand, are difficult to identify since the process adapts to these problems; hence, they are often difficult to correct.

Take the example of purchase requisitions: If requisitions with incorrect budget numbers are occasionally received, this problem can be traced to a new employee who was not given correct instructions in completing forms. The problem can be easily identified and easily corrected by providing instructions. On the other hand, consider the situation in which there is always a backlog in processing requisitions. This occurs because part of the purchasing clerk's job is to call every person who prepared a requisition to find out where the supplies should be delivered. No one may even consider it a problem, even though it could be very costly, because it occurs every time. The backlog problem can be corrected by changing the purchase requisition form and adding a line for "delivery location."

Chronic problems are like high blood pressure. The body adapts and gets used to it. Occasional problems are like a headache. They can be easily identified and cured. A comparison of the two types of problems follows:

	Occasional	Chronic
Occurrence	Infrequent	Frequent
Analysis	Limited data	Broad data
	Simple causes	Complex causes
	Special causes	Common causes
Correction	Localized correction	Wide range of measurements
	Individual action	Management action

The PIT members should go back to each person interviewed who described a problem or an idea to explain what action will be taken. If no action will be taken, the employee needs to be told why. This quick feedback will do a great deal to establish credibility of the PIT and gain future cooperation.

We find that the team will perform better if a problem-tracking list is computerized. This computerized list should include:

- Statement of the problem
- The person who identified it and the date it was identified
- The person who will correct the problem and the date assigned
- The corrective action to be taken and the implementation target date
- Key checkpoints
- The date the corrective action was implemented
- The effectiveness of the corrective action

Reminders should be sent out to each individual 1 week before the due date for the action. The computerized list will also be used as an agenda item at the PIT meeting. This same type of computerized list is a good way to manage the PIT project plan.

Now that the PIT is familiar with all elements of the process, it should look at the whole process to determine the following:

1. Are the boundaries appropriate? If not, have the process owner report the recommended changes to the executive improvement team (EIT).

2. Does the process lend itself to being divided into subprocesses to increase the efficiency of the PIT? If so, the process owner should assign *sub-PITs* to concentrate on these smaller processes. The PIT still should meet to review the total activity to ensure that suboptimization does not occur.

By now, the PIT has collected a great deal of data, and it is time to start a business process project file. This file will house all the collected data and action plans. The first part of the file should provide an overview and include such documents as the PIT mission, goals, primary measurements, and plan. The second part should include all the flowcharts. The third part should include a list of known problems, all action plans, and an assessment of the effectiveness of each action plan. Subsequent parts of the project file will include sections on each activity, copies of all procedures, copies of benchmark studies, etc. This project file should be kept up to date at all times. For complex processes, the project file should be established as soon as the final boundaries are agreed to.

PROCESS EFFECTIVENESS

Process effectiveness is how well the process meets the requirements of its end customers. It measures the quality of the process. More specifically, effectiveness is how well:

- The output of the process meets the requirements of the end customers
- The output of every subprocess meets the input requirements of internal customers
- The inputs from the suppliers meet the requirements of the process

The effectiveness of every process, no matter how well the process is designed, can be improved. Improved effectiveness leads to happier customers and improved sales and market share.

But how do we identify these improvement opportunities? Typically, the individuals involved in the process are too busy with the day-to-day routine to identify and make improvements. In most organizations, employees lack both the authority and the skills to make these improvements. Initially, when the BPI effort is undertaken, the PIT should work with the employees to identify these opportunities. Then, on an ongoing basis, making improvements should be a part of every employee's job. We will discuss the continuous improvement process and how to streamline the process later in the book. For now, let's focus on how the PIT, as it understands the process, can identify these improvement opportunities.

The first step is to select the most important effectiveness characteristics. *Effectiveness characteristics* are indicators of how well the process is functioning. You need to look not only at quality as a whole but also at the subprocesses and major activities within the process. The goal is to be sure that the output meets the customer requirements.

Typical lack-of-effectiveness indicators are:

- Unacceptable product and/or service
- Customer complaints
- High warranty costs
- Decreased market share
- Backlog
- Redoing completed work
- Rejected output
- Late output
- Incomplete output

For the entire process, these are the key measures of effectiveness developed in Chapter 3. For key subprocesses, we may need to identify additional effectiveness characteristics in the same way that we did for the entire process. A comprehensive list of effectiveness characteristics should be created. This list will be created as the walk-through is conducted, based on the interviews.

Next, we should gather information about these effectiveness characteristics. One way to do this is by collecting historical data. How many invoices were rejected during the past 6 months? How many reports were late? How much effort was expended on correcting errors? The purpose of this data is to methodically review the quality of the key activities involved in the process and to try to find the problems (rework, delays, input quality) as well as the potential causes (inputs, methods, training). Be sure to include effectiveness questions as part of your process walk-through questionnaire.

PROCESS EFFICIENCY

Achieving process effectiveness is primarily for the benefit of the customer, but process efficiency is primarily for the benefit of the process owner. Efficiency, as defined in Chapter 3, is output per unit of input (e.g., miles per gallon). Typical efficiency characteristics are:

- Cycle time per unit or transaction
- Resources (dollars, people, space) per unit of output
- True-value-added cost percentage of total process cost
- Poor-quality cost per unit of output
- Wait time per unit or transaction

As you perform the walk-through, look for and record ways to measure efficiency of activities and groups of activities. This data will be used later, when the total measurement process is established.

PROCESS CYCLE TIME

Although cycle time is considered an efficiency measurement, it has a big impact on customers because it affects delivery and cost. *Cycle time* is the total length of time required to complete the entire process. It includes not only the time taken to perform the work but also the time spent moving documents, waiting, storing, reviewing, and reworking. Cycle time is a key issue in almost all critical business processes. Reducing total cycle time frees resources, reduces cost, improves the quality of the output, and can increase sales. For example, if you cut process development cycle time, you may gain sales and market share. If you reduce product cycle time, you will reduce inventory cost and improve delivery. If you reduce the billing cycle, you will have more cash on hand. Cycle time can make the difference between success and failure.

At this point, you should measure the actual cycle time for your process. The actual cycle time probably will be quite different from the theoretical cycle time (processing time) defined in the written procedures or understood by the organization. Often, processing time is less than 1 percent of the cycle time. There are four ways to collect this information:

- End-point measurements
- Controlled experiments
- Historical research
- Scientific analysis

End-Point Measurements

Many processes lend themselves to end-point measurements, particularly repetitive processes that start with a written, dated input and finish when the requested output is delivered. In such cases you have:

- A large number of incidents
- Beginning and end dates that can be correlated
- Information that is available from the current data system by reviewing records or sampling at the end of the process

Purchasing is a good example. It begins with a purchase requisition and ends with a material receipt. You can gather information on beginning and end dates from the existing data systems and calculate an average cycle time for the process.

Controlled Experiments

When information on beginning and end dates is not available using the present data system, or when the information cannot be correlated, controlled experiments can provide you with the needed cycle-time data. This involves:

- Selecting a sample
- Introducing the controlled sample into the process
- Gathering data related to the sample

Be careful not to identify the evaluation sample. You don't want it to receive special attention that would invalidate the data. In addition, controlled experiments are appropriate only for repetitive processes with short to medium cycle times (e.g., claims processing, invoicing, and engineering change notices). Often, it is best to divide the process into logical cycle-time segments. These are groups of activities that trigger another activity or other activities. For example, a sales process cycle-time analysis could be divided into four different experiments:

- Cycle time from receipt of call by the telemarketing group to qualifying a lead and notifying sales
- Cycle time from sales being notified until first customer contact
- Cycle time from first customer contact to when the order is received by the order processing department
- Cycle time from receipt of sales order to when it is entered into the computer

Historical Research

Although some processes are repeated only infrequently (e.g., new product introduction), cycle time is very important. In such cases, some historical research may be necessary to obtain dates documenting the beginning and end of these major processes. A good place to look is the old annual strategic operating plans. You may be shocked at some of these cycle times. Incidentally, this is one area in which Asian countries are far ahead of the United States. (The new product development process for Japanese cars is a good example. It is about 60 percent of U.S. cycle time and about 50 percent of the cost.)

Scientific Analysis

If the first three approaches are not applicable, there still is one good approach left. It involves breaking down the process into smaller components and then estimating each component's cycle time. To help with this analysis, use the flow diagram to determine whether there are any subprocesses, or a series of activities, for which information can be gathered using either end-point measurements or controlled experiments. For other operations, use the knowledge of the people performing the work to estimate cycle time. These data should be collected during the walk-through. Combining all the resulting data will allow you to estimate total cycle time. Correctly done, this type of approach has an amazingly small error rate—frequently less than 5 percent.

PROCESSING TIME VERSUS CYCLE TIME

Consider the cycle time of a letter-writing process:

Activity	Processing time (hours)	Cycle time (hours)
1. A manager estimates that it takes 12 minutes to write a one-page memo and place it in the outgoing basket.	0.2	0.2
2. The manager's secretary picks up outgoing mail twice a day, at 9 a.m. and 1 p.m. Average delay time is 12 hours.	0.1	12.0
3. The secretary assists three managers, answers phones, schedules meetings, processes incoming mail, retypes letters, and performs special assignments. All these activities have priority over typing. Average time before starting to type a letter is 26 hours.		26.0
4. The secretary types the memo and puts it into the manager's incoming mail.	0.3	0.3
5. Incoming mail and signature requests are delivered twice a day, at 9 a.m. and 1 p.m.		12.0
6. The manager reads incoming mail at 5 p.m.	0.1	17.0
7. The secretary picks up mail at 9 a.m.		16.0
8. Retyping is a priority activity and is returned in the 1 p.m. mail delivery (Note: 60 percent of all letters are changed by managers.)	0.2	4.0
9. The manager reads and signs letters at 5 p.m.	0.1	4.0
10. The memo is picked up by the secretary at 9 a.m.	0.1	16.0
11. The memo is put in the copy file and held for the next trip to copy center at 2 p.m.	0.1	5.0
12. The secretary walks to the copy center, makes copies, and addresses envelopes.	0.1	0.1
13. The secretary takes memo to mailbox by 5 p.m.	0.2	2.5
14. Mail is picked up at 8 a.m.		15.0
15. The memo is held in the mailroom for afternoon mail delivery at 3 p.m.	0.1	17.0
16. Secretary 2 picks up mail and sorts at 4 p.m. Secretary puts memo in manager's incoming mail.	0.1	1.0
17. Secretary delivers incoming mail to manager 2 at 9 a.m.	0.1	14.0
18. Manager 2 reads mail at 5 p.m.	0.1	8.0
19. Manager 2 drafts answer and puts it into the outgoing mailbox, telling manager 1 to supply more information. It is classified "rush" because it is now overdue.	0.3	0.3
TOTAL	2.2	170.4

This scenario is closer to truth than fiction. While the process cycle time is 170.4 hours (over 7 working days, or 9 calendar days), only 2.2 hours were spent in actual work effort. The rest was wasted time. It is easy to see why we need to measure cycle time.

COST

Cost is another important aspect of the process. Most organizations divide their financial information by department—because that is tradition. However, as we noted earlier, work flows across departments. Consequently, it is often impossible to determine the cost of the whole process.

The cost of a process, like cycle time, provides tremendous insights into process problems and inefficiencies. It is acceptable to use approximate costs, estimated by using current financial information. Obtaining accurate costs may require an enormous amount of work, without much additional benefit.

The PIT should estimate the cost of the entire process. First, identify all departments involved in the process by reviewing the flowchart. What are the activities? Which departments are primarily responsible for the activity? Which departments support this activity? There often are a number of surprises. The PIT members should work with the managers of each department to get their estimates of how much time the departments spend on the process. These managers may already collect such information on time sheets or similar documents, or at least they can make good estimates. Include the variable overhead cost in all estimates. *Variable overhead* is the overhead that could be eliminated if an activity was eliminated (e.g., floor space, heat, retirement funds). Ask the financial department to provide the variable overhead figure for each organization.

Another way to get a process cost estimate is to obtain a department's total monthly costs from the financial records and then have the department's manager allocate the costs to the process, using the time estimates. Consider an office supplies purchasing process, involving purchasing, shipping and receiving, and accounting departments:

Department	Pur- chasing	Shipping and re- ceiving	Ac- counting	Total
A. Department cost ($/month)	$46,350	$18,500	$75,600	
B. Time spent on process	35%	15%	10%	
C. Direct process cost/month (A × B)	$16,222	$ 2,775	$ 7,560	$26,557
D. Variable overhead cost per person*	25%	30%	20%	
E. Salary of employees in the process	$16,000	$ 2,775	$ 6,500	
F. Overhead cost (D × E)	$ 4,000	$ 833	$ 1,300	$ 6,133
G. Total process cost ($/month) (C+F)	$20,222	$ 3,608	$ 8,860	$32,690

*Not included in department cost (A).

In this case, the cost of the office supplies purchasing process averages $32,690 per month. This will serve as a useful comparison with other processes and as a benchmark for future improvement efforts.

Typically, you will also need to understand the costs at a more detailed level. What do the major subprocesses cost? What do the key activities cost? What is the cost of each output? Make additional estimates, and complete the form shown in Figure 5.1.

Continue with the example of an office supplies purchasing process to calculate the average cost of a single purchase:

- Identify the major subprocesses or activities, using the flowchart. For example, there are nine activities (e.g., recognize needs, write requisition, review requisition, identify suppliers).
- Calculate the cycle time for each subprocess or activity using the techniques discussed earlier. In this case, we estimated that the total processing time is 1.8 days per purchase, and total cycle time is 36.8 days per purchase.
- Estimate the cost for each activity. This has two components: cost of personnel (including variable overhead) for the 1.8 days per purchase and other costs (e.g., computer systems), if they are significant. These are estimated at $571 for a single purchase, which may be for a $25 item.

We now have a good idea of what the cycle time and costs of the process are. You can depict this information on cost-cycle time charts to determine problem areas on which to work. Cost-cycle time charts (see Figure 5.2) display how a typical purchase of office supplies builds up costs

Activity	Cycle Time (days)			Cost ($ per purchase)		
	Processing	Wait	Total	Personnel	Other	Total
1. Recognize Needs	0.1	1.0	1.1	30		30
2. Write Requisition	0.2	2.0	2.2	56		56
3. Review Requisition	0.1	5.0	5.1	28		28
4. Identify Suppliers	0.6	6.5	7.1	175		175
5. Negotiate Terms	0.2	0.5	0.7	58		58
6. Place Order	0.1	10.5	10.6	26	30	56
7. Receive Materials	0.1	7.5	7.6	26		26
8. Check with Order	0.2	1.0	1.2	54	38	92
9. Deliver to User	0.2	1.0	1.2	50		50
Total	1.8	35.0	36.8	$503	$68	$571

Figure 5.1 Cost-cycle time worksheet.

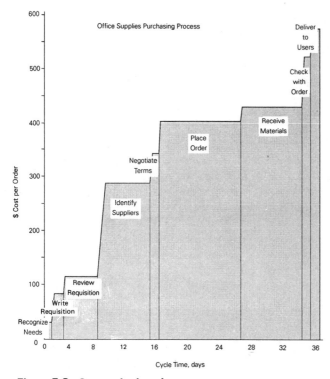

Figure 5.2 Cost-cycle time chart.

over the 36.8 days that it takes from one end of the process to the other—from "recognize needs" to "deliver to user."

In this chart, the horizontal axis represents total cycle time, and the vertical axis represents cost for a single purchase. Upward sloping lines indicate processing time for the activities, while horizontal lines indicate wait time when no direct cost is incurred. If you follow the chart, you can see that:

- The highest cost is incurred to "identify suppliers." Therefore, you should focus on the methods and processes used to identify suppliers.
- There are long wait times, when no activity is being performed, at the "identify suppliers" and "receive materials" stages. Expand these flowcharts to the task level to better understand why they take so much time, and to determine how to improve the process.

You will note in Figure 5.2 that no costs were added to the process cost during the wait periods. Although there is no direct cost added, everything that is put in a hold or wait classification costs the organization money indirectly. It costs for the space it occupies and the delay in final results. Lew Springer, former senior vice president of Campbell Soup, points out that the storage cost for a can of soup is 43 cents on the dollar.

The objective of reviewing cost-cycle time charts is to analyze both the cost and the time components and to find ways to reduce them. This ensures that the effectiveness and efficiency of the process are improved. The next chapter discusses how to streamline the process.

SUMMARY

It is very important for the PIT to take the time to verify that its understanding of the process is truly correct and that all employees are doing the same job the same way. In addition, the PIT needs to understand how the process is performing and where the greatest opportunities are for improvement. This requires the collection of data about process effectiveness, efficiency, cycle time, and costs. Three rules should guide the PIT's improvement effort:

- Do not start until you know what is being done in the process.
- Do not start until all employees are performing the same job in the same way.
- Act on the basis of data, not on the basis of a guess.

Take time at the beginning to understand the process, and collect meaningful data about the process. This will save a lot of wasted effort and reduce the total improvement cycle. A problem well defined is a problem half solved.

6

Streamlining the Process

INTRODUCTION

Up to this point in our text, we have been preparing for the primary goal: the actual improvement of a business process. This chapter will present the remainder of the 10 fundamental business process improvement (BPI) tools, which are grouped under an umbrella called *streamlining*. This term best describes the fundamental concept of improving the business process. It identifies the methods that create positive change in effectiveness, efficiency, and adaptability.

THE PRINCIPLES OF STREAMLINING

Streamlining suggests the trimming of waste and excess, attention to every minute detail that might lead to improved performance and quality. It suggests contouring to provide the smoothest flow, the least resistance to progress and performance with the minimum amount of effort. With streamlining, the process will operate with the least disturbance to its surroundings.

There are 12 cornerstone tools to streamlining, and they are applied in the following order:

1. *Bureaucracy elimination.* Removing unnecessary administrative tasks, approvals, and paperwork.

2. *Duplication elimination.* Removing identical activities that are performed at different parts of the process.

3. *Value-added assessment.* Evaluating every activity in the business process to determine its contribution to meeting customer requirements. Real-value-added activities are the ones that the customers would pay you to do. For example, a customer is willing to pay for the meal served on an airline (real-value-added) but does not care whether you keep records on employees who are on vacation or who come in late.

4. *Simplification.* Reducing the complexity of the process.

5. *Process cycle-time reduction.* Determining ways to compress cycle time to meet or exceed customer expectations and minimize storage costs.

6. *Error proofing.* Making it difficult to do the activity incorrectly.

7. *Upgrading.* Making effective use of capital equipment and the working environment to improve overall performance.

8. *Simple language.* Reducing the complexity of the way we write and talk; making our documents easy to comprehend by all who use them.

9. *Standardization.* Selecting a single way of doing an activity and having all employees do the activity that way all the time.

10. *Supplier partnerships.* The output of the process is highly dependent on the quality of the inputs the process receives. The overall performance of any process improves when its suppliers' input improves.

11. *Big picture improvement.* This technique is used when the first 10 streamlining tools have not provided the desired results. It is designed to help the PIT look for creative ways to drastically change the process.

12. *Automation and/or mechanization.* Applying tools, equipment, and computers to boring, routine activities to free up employees to do more creative activities.

These tools are proven techniques. In fact, some have been so successful in business and industry during the past three decades that they have evolved into entire disciplines. In BPI, they are not viewed as separate methods but are used in concert with each other. You will find that the scope of these tools has been enlarged when applied to the business processes.

Many people in business and industry—especially workers and first-line managers—try to improve their work processes. Their efforts are not insignificant, nor do the people themselves lack motivation. However, many seem to rely on intuition and consequently labor haphazardly, independently, and ineffectively. What progress is made is often too little and too slow compared to the need for continuous improvement.

This chapter presents specific, proven methods of improving the process in a planned and organized way. When you use these methods consistently, your improvement potential and actual accomplishments are multiplied several times over.

Our Definition of Improvement

The word *improvement* seems to have a variety of connotations. Improvement of a process means changing a process to make it more effective, efficient, and adaptable. *What* to change and *how* to change will depend on the particular focus of the PIT team and the process.

The journey to customer satisfaction and beyond will take us through four phases:

Streamlining. The application of basic tools will allow you to make the initial changes to the process.

Preventing. In this stage, you should change the process to ensure that errors never reach the customer. This way, you avoid the correcting phase.

Correcting. If prevention did not work, you should fix what is wrong with the process. In other words, stem the flow of errors. However, it is better to rely on prevention since correcting problems later usually increases cost (covered in Chapter 7).

Excelling. At the end of the correcting phase, the process is satisfactory (i.e., it works, it is stable, and it meets customer requirements). Many companies are willing to stop here, but world-class organizations realize that further improvement is not only possible, but necessary. You can exceed customer requirements and still reduce costs and increase profits. Don't be satisfied with an adequate business process; aim for a process that is competitive and innovative. Learn to *think* and *do* in new ways in order to achieve this (covered in Chapter 8).

When the PIT focuses on improvement activities, it is seeking opportunities to apply the principles of organized and systematic improvement advocated in this book. The PIT must concentrate first on streamlining the process, then on correcting it, and finally on perfecting it.

This sequence may seem out of order. Often, people believe that problem solving should precede streamlining. This may be true if the problems are so great that they are having a major impact on your customer. In this case, put in a quick fix and get right to work on streamlining to eliminate many of the problems. Generally, it is best to make the major process changes and put in the new measurement systems before you start problem-solving activities. The streamlining process will eliminate many of the problems as the process is reorganized, so there is no need to improve a task that soon will be eliminated. When the PIT completes the streamlining phase, the process will be well on the road to excellence. Let's now look, in detail, at each of the streamlining tools and see just how they can be applied to our process.

BUREAUCRACY ELIMINATION

The word *streamlining* suggests the ultimate search for efficiency and effectiveness, an absence of fat and excess baggage, the smooth flow and unrestricted directness of effort and motion. Streamlining implies symmetry, harmony of elements, and beauty of design.

Bureaucracy, on the other hand, means the opposite. It is a major stumbling block to the organized, systemic, companywide implementation of BPI concepts and methods. Bureaucracy is everywhere, even when we don't recognize it. We must learn to actively search for and recognize it. Then, we need to eliminate it.

The Big B in bureaucracy stands for bad, boring, burdensome, and brutal. We often think of bureaucracy as departments with layers of officials striving to advance themselves and their departments by creating useless tasks and rigid, incomprehensible rules. We think of long delays in processing as documents go through multiple channels and levels of review, requiring multiple signatures by people who are never available when needed. Their existence seems to add resistance to progress, adding cost but little real value.

Bureaucracy often creates excessive paperwork in the office. Managers typically spend 40 to 50 percent of their time writing and reading job-related materials; 60 percent of all clerical work is spent on checking, filing, and retrieving information, while only 40 percent is spent on important process-related tasks. This bureaucracy results from organizational or individual personalities caused by such psychological factors as:

- Paranoia about being blamed for errors
- Poor training
- Distrust of anyone
- Lack of work
- Inability to delegate

- Lack of self-worth
- Thrill of checking for and finding minuscule mistakes
- Need to overcontrol
- Unwillingness to share information

The sinister effects of bureaucracy are innumerable and profoundly damaging to every organization and to the BPI effort. Therefore, you should evaluate and minimize all delays, red tape, documentation, reviews, and approvals. If they are not truly necessary, you should eliminate them. A word of caution: Sometimes an activity may not have an obvious purpose but is, in fact, valuable to some other process in the organization, so don't be too quick in your evaluation.

You can identify bureaucracy by asking such key questions as:

- Are there unnecessary checks and balances?
- Does the activity inspect or approve someone else's work?
- Does it require more than one signature?
- Are multiple copies required?
- Are copies stored for no apparent reason?
- Are copies sent to people who do not need the information?
- Are there people or agencies involved that impede the effectiveness and efficiency of the process?
- Is there unnecessary written correspondence?
- Do existing organizational procedures regularly impede the efficient, effective, and timely performance of duties?
- Is someone approving something he or she has already approved? (For example, approving capital equipment that was already approved during the budget cycle.)

The PIT needs to ask questions about each process step and then carefully consider the responses in order to gain insights that will help streamline the process.

Management must lead an attack against the bureaucracy that has crept into the systems controlling the business. Bureaucracy in government and business continues to worsen. These huge paperwork empires must be destroyed if U.S. industry is to flourish. Our copiers are used far too much, and we have too many file cabinets. More than 90 percent of the documents we retain are never used again. Here's an example of bureaucracy's rapid escalation: In 1955, the entire specification for McDonnell Douglas's F-4 airplane was documented on two pages. In contrast, the 1980 proposal for the C-17 airplane consisted of 92 *books* containing 13,516 pages and 35,077 pieces of art. Granted, we need good documentation—but just adding volume does not make it good!

In Boeing's manufacturing division, a team attacked the paperwork bureaucracy and cut six manuals down to one that was smaller than any of the previous six. IBM-Brazil launched a Big B campaign that eliminated 50 unnecessary procedures, 450 forms, and 2.5 million documents a year.

Many activities do not contribute to the content of the process output. They exist primarily for protection or informational purposes, and every effort should be made to minimize these activities.

The PIT is apt to run into resistance because of varying opinions and organizational politics. Overcoming resistance to eliminating bureaucracy takes skill, tact, and considerable planning. Bureaucracy's impact on cost and cycle time should be calculated; its impact on the internal and external customer should be understood. Once the full impact of bureaucracy is understood by all concerned, it is difficult to justify retaining the activity. The entire organization should continually eliminate every example of bureaucracy.

The attack on bureaucracy should start with a directive informing management and employees that the company will not put up with unnecessary bureaucracy, that each approval signature and each review activity will have to be financially justified, that reducing total cycle time is a key business objective, and that any no-value-added activity that delays the process will be targeted for elimination.

After completing the flow diagram, the PIT should review it, using a light blue highlighter to designate all activities related to review, approval, second signature, or inspection. The PIT will soon learn to associate blue on the flowchart with bureaucracy. These blue activities become key targets for elimination.

The managers responsible for each of the Big B activities should justify activity-related costs and delays. Often, a manager will try to push the matter aside, saying, "It only takes me 2 or 3 seconds to sign the document. That doesn't cost the company anything." The answer to a remark like this is, "Well, if you don't read the document, there is no reason to sign it."

In one company that started a Big B elimination campaign, a group of 10 capital equipment requests were processed through five levels of management approval signatures. Two requests contained only a legitimate cover sheet with blank pages attached instead of the written justification required. And these two requests made it through all five levels of approval. This experiment shocked management into fully backing the Big B elimination campaign.

The cost of Big B activities is more than we realize. Reading and approving a purchase order may cost the controller only 1 minute of time, but the processing to obtain that signature costs much more. Let's look a little closer.

	Time (minutes)	Cost (dollars)
1. Manager goes to the supply cabinet to get envelope.	5	6.00
2. Manager looks up controller's office address and addresses envelope.	5	6.00
3. Manager takes envelope to mail drop and returns to office.	7	9.80
4. Mail is picked up and taken to mailroom, sorted, and delivered to controller's secretary.	1800	2.00
5. Secretary logs in mail and puts into controller's incoming mail folder.	4	.60
6. Waiting for controller to read document.	800	0
7. Reading and signing documents.	1	1.20
8. Waiting for secretary to process.	800	0
9. Secretary logs out the document.	2	.30
Total:	3424	25.90

In this example, the controller needs to justify delaying the purchase order an additional 2 days and increasing its cost by $26.00. The real losses caused by Big B activities are always much more than we originally estimate.

The justification for retaining a Big B activity requires some data. How many items are rejected? How much does the company save when an item is rejected? Rejecting an item does not necessarily mean savings for the company; quite the contrary. A rejected document may cause more bureaucracy, more delay, and increased costs. Time-cycle delay costs are based on the advantage to the customer if the output from the process is delivered early. The justification for each Big B activity should be based on the potential loss or gain to the company. If it is a breakeven point, the activity always should be eliminated. Many companies require a 3 to 1 return on any investment. This rule should apply to bureaucracy activities, also. A bureaucracy step should be left in only if there is a sizable, documented savings from the activity. Even then, the PIT should look at why the bureaucracy activity is saving money and see if there is any other, less expensive way of accomplishing the same result.

To purchase a ballpoint pen (or anything else, for that matter) at Intel took 95 administrative steps and 12 pieces of paper. When the company eliminated the bureaucracy, the purchase took eight administrative steps and one form. Intel estimated its attack on Big B improved productivity by 30 percent and saved $60 million a year. It would take the equivalent of $277 million in increased sales for Intel to generate $60 million in profits.

DUPLICATION ELIMINATION

Look at the process. If the same activity is being performed at different parts of the process or by different individuals in the process, see if both activities are needed (for example, two similar databases). Frequently the same, or very similar, information is generated at different parts of the process, often by different organizations. Not only does this add to the overall cost of the process, it also provides the possibility of having conflicting data that unbalance the process. Frequently, a department within the process will generate information, and a supplier will generate similar information, providing it to a different department. (For example, purchasing may provide the price for a new piece of equipment, and industrial engineering may get a different price from an equipment supplier. Sales may generate a monthly customer product ship forecast, and production control distributes a completely different forecast.) Because management has not trusted the information system, many personal processes have been set up to provide checks and balances. Management keeps its own absentee records, and the employee relations department generates another report. In today's competitive environment, we cannot afford these duplications and the turmoil created when there is a difference in the two data sources. Data integrity is key to our business processes. We cannot afford duplicate data sources to check on each other. What we need to do is build integrity into a single source.

In other cases, we have redundancy because the work groups do not realize that the activity has already been performed, or the process has not been designed to link the using organizations to the earlier output. This again provides an opportunity to improve the overall organization's effectiveness.

VALUE-ADDED ASSESSMENT

Value-added assessment (VAA) is an essential principle in the streamlining process. The technique is simple and direct and very effective. To understand this tool's importance, first explore the concept of value added through this simplified analogy of a product's manufacture.

When raw materials, subassemblies, or other substrata materials progress through a manufacturing process, they accumulate what could be called *value added*. Companies typically apply a bookkeeping method of tracking value added by apportioning costs incurred during production. This is called *accrual*. The goal is to be sure that the value of the end product (accrued value, usually measured in dollars) exceeds the accumulated costs.

On the accounting books, value added essentially reflects a theoretical increase in value over and above the original cost. This is usually assumed to be greater than the accumulated costs that have been "added" along each stage of the production process. It is theoretical as far as it applies to both market value (dependent on what customers are willing to pay) and cost value (dependent on the bookkeeping method used). Value added, then, is the value after processing, minus the value before processing. A mathematical representation might look like this:

$$VA = V_2 - V_1$$

where VA = value added
   V_2 = value after processing
   V_1 = value before processing

There are many subjective issues that relate to value (e.g., utility value, aesthetic value, prestige value, and cross-impact value). These qualities sometimes have as much importance as economic value, but in most cases, monetary values must be carefully considered.

It is not necessary to understand this idea of value added in great detail, or even to understand all the implications of the word *value*. PIT members need to understand that each step of a process involves a cost to the enterprise (e.g., labor, overhead, materials, storage, or transportation). At each step of the process, costs accrue. These will be considered for book value, regardless of the real value. But the organization's goal should be to ensure that every activity contributes real value added as far as is possible. Ideally, this should be equal to or greater than the actual costs incurred. We must also recognize that the customer's perception of value is independent of the actual costs incurred to provide the product or service.

Real-value-added (RVA) activities are those activities that, when viewed by the end customer, are required to provide the output that the customer is expecting. There are many activities performed that are required by the business but that add no value from the customer's vantage point (business-value-added, or BVA, activities). In addition, there are many activities that add no value, for example, storage.

Value-added assessment (VAA) is an analysis of every activity in the business process to determine its contribution to meeting end-customer expectations. The object of VAA is to optimize BVA activities and minimize or eliminate no-value-added activities. The organization should ensure that every activity within the business process contributes real value to the entire process.

Value is defined from the point of view of the end customer or the business process. Activities that must be performed to meet customer requirements are considered RVA activities. Activities that do not contribute to meeting customer requirements, and could be eliminated without degrading the product or service functionality or the business, are considered no-value-added (NVA) activities. This definition of value is independent of the costs incurred to provide the activity.

Every activity within the process (or subprocess) should be reviewed by the PIT, and an assessment should be made: Does this activity add value to the end customer or the business? There are two kinds of NVA activities:

- Activities that exist because the process is inadequately designed or the process is not functioning as designed. This includes moving, waiting, setting up for an activity, storing, and doing work over. These activities would be unnecessary to produce the output of the process but occur because of poor process design. Such activities are often referred to as part of poor-quality cost.
- Activities not required by the customer or the process and activity that could be eliminated without affecting the output to the customer (for example, logging in a document).

Figure 6.1 shows how the evaluation is done. RVA activities contribute directly to producing the output required by the end customer. The cost and cycle time of each activity (collected in the process documentation phase of Chapter 4) can be analyzed for value added versus no value added. The PIT should analyze each activity and/or task on the flowchart and classify it as an RVA, a BVA, or an NVA activity.

Use a yellow highlighter to designate each BVA activity. Color in the NVA activities with a pink highlighter. You have now turned your flowchart into a rainbow flowchart. Typically, as PIT members go through this phase of the analysis, they are astonished at the small percentage of costs that are RVA activities. For most business processes, less than 30 percent of the *cost* is in RVA activities. Even more alarming is the mismatch of cycle time of RVA activities compared to total cycle time. For most business processes, less than 5 percent of *time* is spent in RVA activities.

Obviously, this indicates something very wrong, and managers are often disturbed when they learn of these numbers. But there are several explanations:

- As the organization grows, processes break down and are patched for use, thereby making them complex.
- When errors take place, additional controls are put in place to review outputs rather than change the process. Even when the process is corrected, the controls often remain.

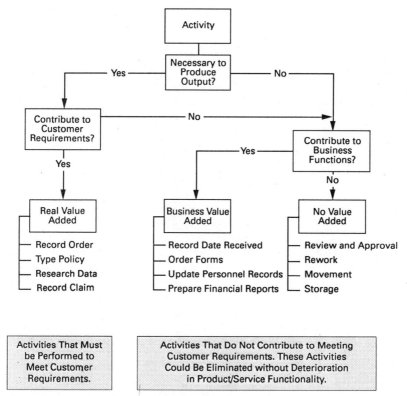

Figure 6.1 Value-added assessment.

- Individuals in the process seldom talk to their customers and hence do not clearly understand the customers' requirements.
- Too much time is spent on internal maintenance activities (such as coordinating, expediting, record keeping) instead of on redesigning the process.

Once the value and cycle time of all activities have been determined, the PIT should record this analysis on a cost-cycle time chart, as discussed in Chapter 5 (see Figure 5.2). Figure 6.2 is an example of how a cost-cycle time chart will look when no-value-added and business-value-added activities and/or tasks are highlighted. The PIT should now answer the following questions.

- How can the RVA activities be optimized?
- Can the RVA activities be done at a lower cost with a shorter cycle time?
- How can the NVA activities be eliminated? If they cannot, can they be minimized?

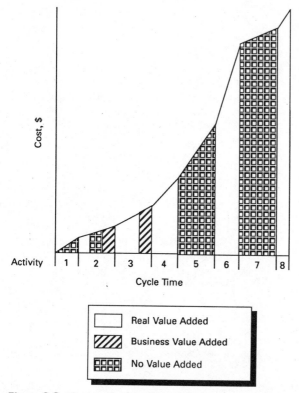

Figure 6.2 Cost-cycle time chart.

- Why do we need the BVA activities? Can we minimize their cost and cycle time?

The PIT has to be very creative in coming up with solutions and should not be constrained by the current culture, personalities, or environment.

- Rework can be eliminated only by removing the causes of the errors.
- Moving documents and information can be minimized by combining operations, moving people closer together, or automation.
- Waiting time can be minimized by combining operations, balancing work loads, or automation.
- Expediting and troubleshooting can be reduced only by identifying and eliminating the root causes.
- NVA outputs can be eliminated if management agrees.

- Reviews and approval can be eliminated by changes in policies and procedures.

Challenge everything. There is no sacred cow in BPI. Every activity can always be done in a better way. The end result of this analysis is an increase in the proportion of RVA activities, a decrease in the proportion of BVA activities, a minimizing of NVA activities, and a greatly reduced cycle time (see Figure 6.3). This concept is so important that all employees should learn to use it in their daily work. The results will be powerful.

SIMPLIFICATION

Simplification is another important concept in streamlining. It is similar to the well-known concept of *work simplification*. Let's begin by trying to understand the term. We live in a world of ever-present and increasing

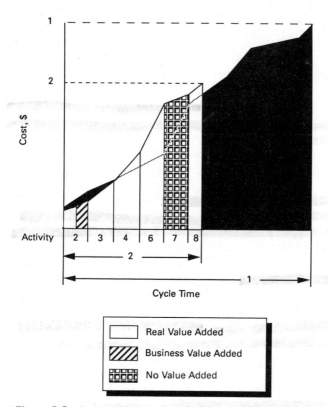

Figure 6.3   Cost-cycle time chart before and after applying VAA.

complexity. Complexity means that life has more of everything: more parts, more systems, more relationships, more dependencies, more problems, and more imperatives. This applies in particular to business processes. Business processes are typically designed to support goals, requirements, and existing volume. However, goals, requirements, and volume are changing, so the processes adapt accordingly. More steps, more tasks, more people, more interdependencies are added. When new tasks are added, support tasks usually follow (for example, preparation, filing, or putting away work), making the process more complex.

The increase in complexity results in increasing difficulties everywhere as activities, decisions, relationships, and essential information become more difficult to understand and more difficult to manage. In an era of rapidly increasing complexity, it is essential to actively and continuously be engaged in simplification as a counterforce to evolving complexity.

So what does simplification mean? It means to reduce complexity wherever feasible. It leads to fewer stages, fewer tasks, fewer interdependencies, etc. It means making everything easier: easier to learn, easier to do, easier to understand.

When you apply simplification to business processes, you evaluate every element in an effort to make it less complex, easier, and less demanding of other elements. When an organization fails to make continuous simplification efforts a major part of the managing process, it invites difficulty and poor performance. The natural evolution of complexity will eventually stifle the ability to manage the system and the processes effectively.

The following list illustrates the application of the concept in relatively simple but time-consuming everyday activities:

- *Duplication and/or fragmentation of tasks.* This can be managed by identifying duplication and fragmentation that occur during various parts of the process and then combining related tasks and eliminating redundancies. Most companies are shocked at the savings possible in this area.
- *Complex flows and bottlenecks.* These can be managed by changing the order of tasks, combining or separating tasks, and even balancing the work load of different individuals.
- *Memos and other correspondence.* These can be simplified by making them shorter, more direct, better formatted, more readable. Thousands of employee hours are saved by decreasing time writing, reading, and interpreting the written word. Less rework is required because of improvements in understanding.
- *Meetings.* An agenda (sent well in advance) is a basic simplification

device. Presentation materials should be simple and easily understood. Meeting protocol should be established and followed, and meeting attendees should be trained in protocol. Fewer meetings, and less time spent in each meeting, are best. Don't schedule meetings in whole increments (1 or 2 hours). In a 1-hour meeting, 80 percent of the work gets done in the last 15 minutes. Look at the meeting agenda schedule to determine the amount of time required. Some meetings should be 25 minutes, others, 80 minutes. Too many meetings are scheduled by the hour (8:00 a.m. to 9:00 a.m.). Try to schedule your meetings by the minute (8:18 a.m. to 8:52 a.m.). Then always start them on schedule. It will make a big difference in the way the team reacts to your meetings.

- *Combine similar activities.* Can similar or consecutive activities be combined to make one job more rewarding to the person performing the assignment and reduce cost, errors, and cycle time?
- *Reduce amount of handling.* Can you reduce the amount of handling by combining responsibilities? Can the person doing the activity evaluate the output to ensure that it is correct? Can a phone call eliminate the need to mail a document to another building? Can a list of documents processed replace copies of the documents that are mailed?
- *Eliminate unused data.* Do you use all the data that are recorded? If not, why record them? Each piece should be challenged.
- *Eliminate copies.* Are all the copies of letters and computer reports used? In most cases, they are not. Every 6 months, you should question the usefulness of all regularly scheduled reports. Send out a letter notifying the recipients that they will be eliminated from the mailing list unless they request otherwise in writing. Tell them that the cost of generating the report will be shared equally among the people who receive copies. You will be surprised how many reports are not needed.
- *Refine standard reports.* Meet frequently with the people who receive standard reports to find out what parts of the report they use and how they use them. Put all the standard reports in similar formats. This reduces the time to read the reports and reduces errors in the interpretation of the reports. When graphs are used, draw an arrow indicating which direction is good. Remove unused parts of the reports.

Let's look at an example of the simplification concept: writing checks, recording the transactions in a journal, and tracking the receipts. With the traditional, old-fashioned manual method, these are three separate activities, tedious and time consuming. Simplification combines them into one activity. The method is called the *one-write system* of check writing. Simultaneously, you can write a check, make a duplicate, and record the transaction in a journal. This system accomplishes the same objective with less effort, less time, and less chance for error. Also, keeping track of receipts is easier because they are attached to a duplicate check that can be filed numerically. Coupling this to a computer database makes the whole process even more effective. This method has proved to be an important part of the process, yet it is so basic.

In trying to find ways to apply the principles of simplification, we would again begin by asking questions such as:

- Is this process effectively systematized or performed haphazardly?
- Would a different process be more effective, more efficient?
- Would a different layout make work smoother and easier, with less handling and less wasted motion?
- Can the forms be filled out without adding another document?
- Do people make errors in filling out the forms?
- Can this activity or stage of the process be eliminated?
- Can this activity or stage be combined with another?
- Could a single activity produce a combined output?
- Are instructions immediately available, easy to understand, self-explanatory?
- Would a backup process eliminate rework or wait time?
- Does this activity require someone to stand by idly while the task is being done?
- Would simpler language speed up reading, improve understanding?
- Does the way it is done create more unnecessary work downstream?
- Is time lost looking for information or documents?
- Could one-time-only serve all?
- Do interruptions of the work flow add to complexity?
- Could a template be used to simplify performing the activity?
- Does the work flow smoothly around the area?
- Is there unnecessary movement?

And the list goes on and on....

Sometimes it is helpful to start your simplification analysis for an activity by asking the question, What is your output? Then design a process for the simplest way of generating that output and compare this new process design to the original process. After the comparison has been made, combine the two processes, taking the best of each.

PROCESS CYCLE-TIME REDUCTION

Critical business processes should follow the rule of thumb that time is money. Undoubtedly, process time uses valuable resources. Long cycle times prevent product delivery to our customers and increase storage costs. A big advantage Japanese auto companies have over American companies is their ability to bring a new design to the market in half the time and cost. Every product has a market window. Missing the early part of the product window has a major impact on the business. Not only does the company lose a lot of sales opportunities, but it is facing an uphill battle against an already established competitor. With the importance of meeting product windows, you would think that development schedules would always be met. Actually, few development projects adhere to their original schedule.

The object of this activity is to reduce cycle time. This is accomplished by focusing the PIT's attention on activities with long real-time cycles and those activities that slow down the process. The time-line flowchart provides valuable assistance in identifying the focus activities. The team should look at the present process to determine why schedules and commitments are missed, then reestablish priorities to eliminate these slippages, and then look for ways to reduce the total cycle time. Typical ways to reduce cycle time are:

- *Serial versus parallel activities.* Often, activities that were done serially can be done in parallel, reducing the cycle time by as much as 80 percent. Engineering change review is a good example. In the old process, the change folder went to manufacturing engineering, to manufacturing, to field services, to purchasing, and finally, to quality assurance for review and sign-off. It took an average of 2 days to do the review in each area and 1 additional day to transport the document to the next reviewer. The engineering change cycle took 15 working days, or 3 weeks, to complete. If any one of the reviewers had a question that resulted in a change to the document, the process was repeated.

By using computer-aided design (CAD), all parties can review the document simultaneously and eliminate the transportation time. This parallel review reduces the cycle time to 2 days. Another less equipment-intensive approach would be to hold weekly change meetings. This would reduce the average time cycle to 3.5 days and eliminate most of the recycling, because the questions would be resolved during the meeting.

- *Change activity sequence.* The geographic flowchart is a big help to this activity. Often, output moves to one building and then returns to the original building. Documents move back and forth between departments within the same building. In this stage, the sequence of activities is examined to determine whether a change would reduce cycle time. Is it possible to get all the signatures from the same building before the document is moved to another location? When a document is put on hold waiting for additional data, is there anyone else who could be using the document, thereby saving cycle time later on?

■ *Reduce interruption.* The critical business process activities should get priority. Often, less important interruptions delay them. People working on critical business processes should not be located in high-traffic areas, such as near the coffee machine. Their phones should be answered by someone else. The office layout should allow them to leave their work out during breaks, lunch, or at day's end. The employee and the manager should agree on a time when the employee will work uninterrupted, and the manager should help keep these hours sacred.

■ *Improved timing.* Analyze how the output is used to see how cycle time can be reduced. If the mail pickup is at 10:00 a.m., all outgoing mail should be processed before 9:45 a.m. If the computer processes a weekly report at 10:00 p.m. on Thursday, be sure that all Thursday first-shift data are input by 8:00 p.m. If you miss the report analysis window, you may have to wait 7 more days before you receive an accurate report. If a manager reads mail after work, be sure that all of that day's mail is in his or her incoming box by 4:30 p.m. It will save 24 hours in the total cycle time. Proper timing can save many days in total cycle time.

■ *Reduce output movement.* Are the files close to the accountants? Does the secretary have to get up to put a letter in the mailbox? Are employees who work together located together? For example, are the quality, development, and manufacturing engineers located side by side when they are working on the same project, or are they located close to other people in the same discipline?

■ *Location analysis.* Ask the question, Is the process being performed in the right building, city, state, or even the right country? Where the activity is performed physically can have a major impact on many factors. Among them are:

   Cycle time
   Labor cost
   Customer relations
   Government controls and regulations
   Transportation cost
   Employee skill levels

■ Performing the activity in less than the optimum location can cause problems, from a minor inconvenience all the way to losing customers and valuable employees. The approach and consideration for selecting the optimum location vary greatly from process to process. As a general rule, the closer the process is located to the customer, the better. The restraints to having the process and its customer in close proximity are economy of scale, stocking costs, equipment costs, and utilization considerations. With today's advances in communication and computer systems, the trend is to go to many smaller locations located either close to the supplier or close to the customers. Even the large manufacturing specialty departments (machine shop, welding department, toolroom, etc.) are being separated into small work cells that are organized to fit a process in which a lot size of one is the production plan. Often, the advantages of quick response to customer requests, increased turns per year, and decreased inventories far offset the decreased utilization costs.

Questions like, Should we have a centralized service department or many remote ones? require very careful analysis. A graphic flowchart helps make these decisions, but the final decision must be based on a detailed understanding of customer expectations, customer impact, and cost comparisons between the options.

- *Set priorities.* Management must set proper priorities, communicate them to employees, and then follow up to ensure that these priorities are lived up to. It is often a big temptation to first complete the simple little jobs—the ones that a friend wants worked on, the ones someone called about—and let the important ones slip. It's the old "squeaky wheel" message. As a result, projects slip, money is lost, and other activities are delayed. Set priorities and live by them.

You will note that most streamlining methods improve cycle time, not just the ones listed in this section of the chapter. This is necessary because cycle time is so important to every organization and to most customers.

ERROR PROOFING

We all have hundreds of opportunities every day to make errors, and they are easy to make. We could drive right through the garage. We could kiss the dog good night and put out the wife, but we don't. It's amazing to think about how well we perform. But we still make errors. We get up and put on two different socks. We sign our name on the wrong line of a document. We estimate we can get a job done quicker than we do. We address envelopes and put the wrong letter inside. The list goes on and on. Yes, there are lots of opportunities for all of us to make errors, and they are so easy to make. Distractions cause errors. We answer the phone and forget that we have not put all the attachments in an envelope we are mailing. Someone asks a question, and we skip a number in the log book. We feel pressure to get out work and take shortcuts that cause errors. As a result, almost nothing is truly error-free. It is so easy to make an error. What we have to do is make it difficult to create errors. There are thousands of ways to do that; the quantity is limited only by our imagination.

Typical error-proofing methods are:

- Put all letters in envelopes with plastic windows to reveal the name and address. Not only does this save much typing time, it also eliminates letters being sent to the wrong person.
- Use different-colored paper for different jobs to help direct correspondence to the right location. Different-colored paper for each day quickly tells everyone what needs to be done today. Different-colored folders for different jobs give the same message. Proper use of colors will greatly reduce errors.
- Put confidential information on paper preprinted with the words "Do not copy" in large thin letters on each page. This is more effective than typing "Confidential" on the page but do both if it makes you feel more comfortable.
- Use preprinted carbon copy lists for repetitive mailing. It is more accurate to delete names for one-of-a-kind mailing than to retype the names.
- Use longer paper if you want a document to be given special care. If you normally use 8½ × 11, go to legal-size paper so that the document will stand out.
- Use computer programs that check spelling, and check the input to ensure that letters, not numbers, and vice versa, are recorded in the correct places on a form. This double-check saves a lot of errors.
- Make sure the on/off switch is out of the way on your computer so that you can't turn it off in error and lose your data.
- Select a phone without a disconnect button. Too often, employees hit the disconnect button instead of the hold button, and a customer is lost.
- For often-used slide shows, mark the slides when they are loaded correctly in the slide holder. Run the mark slowly across the group of slides so that you can easily see if one slide is out of order or upside down.
- Ensure effective communication by asking employees to repeat instructions to be sure they are understood.
- Write down any directions to employees for their future reference.
- Use cross-checking when totaling a number of columns.

You will find it useful to ask yourself, "If I wanted to do this job wrong, how would I do it?" Make a list of the things you could do wrong and then use error-proofing methods to eliminate or minimize the possibility of making an error. This is called *negative analysis*.

UPGRADING

People around the world know and understand the importance of upgrading the equipment used on the manufacturing floor. In the business process as well as the manufacturing process, equipment and environment upgrading is very important. An old typewriter whose keys stick can cause more problems than the old lathe in the machine shop.

Think about it. Is your copier up and running, or is it down half the time? Does it copy on both sides? If so, are your employees instructed to copy all documents on both sides so that filing space will be minimized? Will the copier staple the documents together automatically? Are headsets to free up both hands available for people who use the phone a lot? Do salespeople use cellular phones to keep in close contact with customers? Have you simplified your filing and document storage system by using microfilm? Is bar coding used when appropriate? Are there cycling file bins that put the correct document at the employees' fingertips? Are you using quick-dial phones that save a lot of time and reduce errors on frequently called numbers? An advanced phone system with features like call forwarding, voice mail, and digital readout is a must in most businesses today. Have you purchased the latest computer programs to help your employees become more effective? Do all your key people have and use pagers?

Look at your office layout. Does it meet a strict office standard rather than employee needs? The same desk doesn't work well for all jobs. The programmer who works with lots of computer printouts needs a different office layout and different equipment than the product engineer. If the employee uses a personal computer, does the desk layout make it easy to use? Do your office chairs fit the people and the job, or is there just one style available that makes everyone just a little uncomfortable? Customized office equipment gives the individual a feeling of worth and helps increase efficiency and effectiveness. Are video-training centers located close to the employees who need to be trained? Are conference rooms readily available for meetings? Are colored graphics used to help present important information in a memorable way? Are the conference rooms equipped with "write boards" that automatically transfer the written information from the board to paper?

In reality, the office itself is part of the equipment you provide to the employee. There are many factors you should consider establishing in the office environment. Boxed-in offices discourage communication. Keep your offices open and airy to encourage free exchange of information. Put your heavy-duty printers and copy machines in an enclosed office on the outside of the building so that the noise does not distract your employees. Eliminate the speaker-type paging system. It is effective at finding people, but the interruption to the total population cannot be justified in most cases.

Lighting and color scheme are important office considerations. A shabby, drab, dark office often contributes to worker errors. Bright, cheerful offices produce bright, cheerful people. Most white-collar work depends on use of the eyes. Eyeglasses and white-collar work go hand in hand. A well-lit office minimizes eye strain. Keep the office clean and neat. Paint and redecorate frequently. Have teams of employees select the colors. After all, they are the ones who have to live with it.

Don't forget the most important upgrading of all—upgrading people, often called *technical vitality*. Training and education are an investment in your people and your organization that pay big dividends in improved loyalty and performance. Education maintains your people's competency, just as grease and oil are used to maintain your car. Your best competitive edge is a well-trained work force. Without it, an organization cannot be world-class.

SIMPLE LANGUAGE

Most business writing today cannot be read or understood easily. After a cursory glance, it usually is routed to the nearest file cabinet or trash can. While today's fast-paced world makes receiving accurate and timely information more important than ever, the quality of most business writing lingers in the dark ages. It is pompous, wordy, indirect, vague, and complex. One critic noted: "Too often, business reports are wanting in everything but size."

The PIT needs to evaluate the present documents used in the process to ensure that they are written for the user.

Here are some key factors that will help you simplify communications:

- Determine the reading and comprehension level of your audience. The document should be written so that all readers can easily comprehend the message. If one reader has an eighth-grade comprehension level, prepare the document for seventh-grade. Because your audience all graduated from high school, don't assume that they can read and comprehend at the twelfth-grade level. Many college graduates' reading and comprehension levels are below the tenth grade; when English is a second language, the reading and comprehension level often is much lower than the general education level. When writing for people whose second language is English, write at a level three grades below their general education level, and use the dictionary's first preference meaning only.

- How familiar is the audience with the terms and abbreviations? Unless it is critical to the work assignment, don't use new terms and jargon. If it is necessary to use these words, be sure to clearly define them.

- All procedures more than four pages long should start with a flowchart containing annotations that lead the reader to the detailed paragraph within the procedure.

- Use acronyms with care. It is better to repeat the phrase and take a little more space than to use a shortened version (for example, total quality management—TQM, or business process improvement—BPI). Don't force the reader to learn a new acronym unless it will be used frequently throughout the document. Never use an abbreviation unless it is defined in the document.

It doesn't have to be long to be good. Just think about the following documents.

Document	Number of words
Lord's Prayer	57
Ten Commandments	71
Gettysburg Address	266
Declaration of Independence	300
U.S. Government Contractor Management System Evaluation Program	38,000

B. A. Hardesty, 1985 Streamlining/Tailoring Conference, Los Angeles, CA.

Forms

Generally, we do not give enough thought to forms when we are developing them. Much needless effort is expended, and many errors are created, because forms are poorly designed. Good form design requires a lot of thought. The form should be self-explanatory. Information should be recorded only once. All abbreviations must be defined on the form.

Does a good form make a difference? When British government agencies focused their attention on form design, errors plummeted and productivity soared. For example:

- The British Department of Defense redesigned its travel expense form. The new form cut errors by 50 percent, the time required to fill it out by 10 percent, and processing time by 15 percent.
- By redesigning the application form for legal aid, the British Department of Social Security saved more than 2 million hours per year in processing time.

STANDARDIZATION

A specific task for a PIT to address is the accuracy and adequacy of the documentation covering the process. Business processes frequently are not as well documented as production processes. A production process typically will include such documentation as:

- A set of blueprints specifying all dimensions and assembly of the product
- Detailed specifications covering every chemical or other process solutions applied to the product
- Step-by-step "work instructions" spelling out exactly in what sequence each operation on the product is to be performed, on which machine, and in which department
- A set of inspection and/or test instructions specifying at which specific points in production the product is to be inspected and/or tested, how that inspection and/or test is to be carried out, and the criteria for acceptance or rejection
- A training plan defining background skills and specific task-related training that must be provided to employees before they are qualified to perform the activity

Standardization of work procedures is important to ensure that all current and future employees use the best ways to perform activities related to the process. When each person is doing the activity differently, it is difficult, if not impossible, to make major improvements in the process. Standardization is one of the first steps in improving any process. This is accomplished by the use of procedures.

Procedures should exist for performing most activities. Procedures tell management and employees how the process functions and how to do the activities. These procedures should:

- Be realistic, based on careful analysis
- Clarify responsibilities
- Establish limits of authority
- Cover emergency situations
- Not be open to different interpretations
- Be easy to understand
- Explain each document, its purpose, and its use
- Define training requirements
- Define minimum performance standards

Often, procedures include a flowchart, in addition to written instructions.

All employees should receive copies and then should be trained in the procedures. Unless these methods are communicated, they are of no use. Also, the procedures should be checked and updated regularly.

While it is not the goal of the PIT to create an unnecessary administrative pile of paperwork, there is one best way of performing a task, and that should be the "standard." Standards also set limits of authority and responsibility and must be communicated to the employees. Standardization requires documentation showing how the process is to be carried out, what training of personnel is required, and what is acceptable performance.

SUPPLIER PARTNERSHIPS

All processes are highly dependent on people outside the process who provide input in the form of materials, information, and/or ideas. Before we go any further, we need to examine each process input to see:

- Does the process really need the input?
- Is it entering at the right place?
- Is it at the right quality level?
- Is the timing correct?
- Is it received in the best possible format?
- Do you get more than you need?

Just as your process is a supplier of products and/or services to your customer, people who provide input into your process are your suppliers. In these supplier-customer relationships, both parties have responsibilities. The customer (you) has the responsibility to provide the supplier with documented input specifications that define needs and expectations. The supplier should carefully review the specifications and agree that they can be met. If they cannot, you need to work with the supplier to understand what can be supplied and help the supplier develop a plan to upgrade his or her output if necessary.

The customer should never ask for more than is needed and will be used. Remember, nothing is free. Everything costs the organization that created it something. The process may not pay for it directly, but it costs someone, somewhere, time and effort. Never ask for more than you will use. It is wasteful. In addition, the customer has the responsibility to provide all suppliers with feedback that measures the suppliers' ongoing performance against the agreed-to requirements.

Question whether the input is really needed. Talk with the people who receive the input. See how they use it. Ask what would happen if they didn't receive the input. If you do not pay directly for the input, go to the supplier and ask how much it costs to generate the input. Then compare these costs to the value added to the process by these inputs. Try to eliminate as many inputs to the process as possible. Each input represents a potential delay to your process, additional problems, and additional costs. Frequently, people get data because they think the data are being generated for another use, so there is very little additional cost to add them to the carbon copy list. Often, reports are justified based on the number of people who get them, not on the number of people who use them.

Is the input entering the process at the right place? Can the input enter the process at a different point that would make the process more effective and/or simplify the process? Can it be brought in later to save storage, to make the input more correct, and/or to allow the supplier more time to prepare his or her output?

Are you receiving the input in the best format? There can be a lot of time lost because of the way input is delivered to the process. Often, input data need to be copied from one form to another or input from a computer list into another computer. Numerous phone calls are made because no one answers the phone or the right people are not available. Valuable messages are miscommunicated. Time is lost because large reports need to be analyzed to find the required information. Frequently, information can be communicated by computer disk or over telephone lines much better than in a printed document. The supplier should deliver a product in the most usable form for the customer.

The other half of the partnership is the supplier. The supplier has the responsibility for delivering on-time products and services to the process that meet or exceed the documented specification. In addition, the suppliers should continuously strive to provide these inputs at reduced cost and higher quality levels. The suppliers have the responsibility to really understand how their inputs are utilized. Often, the supplier can make a slight change in his or her input that will have a major, positive impact on the business process.

BIG PICTURE IMPROVEMENT

Although this chapter has focused on making incremental changes to the present process, there are times when these approaches do not bring about the desired results. This normally occurs when the process has already been improved and there is little to gain from further refinement. In these cases, it is time to step back and look at the "big picture."

The big picture technique requires the PIT to step out of today's process and define what the perfect process would be without the constraints of the present organization and/or process. The advantages of this approach are:

- It gives the PIT a way to create new concepts.
- It provides the PIT with a new view of the process.
- It allows the PIT to focus on key success factors.
- It allows the PIT to develop new options.
- It allows the PIT to get around current organizational barriers.

The big picture improvement technique is an effective way to bring about a substantial change in the way we do business.

AUTOMATION AND/OR MECHANIZATION*

In streamlining the process, you probably have already observed opportunities to apply office automation, since most organizations already use computers for at least some business functions. Further, it is a good idea to start thinking early in your BPI analysis about automation, because the best computer systems result from long-term planning. But the point is: Don't switch from manual operation before you understand it well enough to have it working smoothly. And don't introduce more sophisticated automation until you thoroughly analyze the strengths and weaknesses of your existing system. Always keep in mind that every computer system is fully capable of producing, at superhuman speed, both brilliant guidance and business garbage. Automating a mess just produces a faster mess.

As with other aspects of BPI, automation should be under the control of the business process owner, with the full collaboration of the PIT. Effective automation of an operation requires teamwork by people throughout the organization. As an example of how to carry this off well, consider a success at Ford Motor Company.

A few years back, Ford's procurement costs were especially high because of the large number of employees responsible for issuing payments to suppliers. After spotting this as an area for improvement and analyzing alternatives, Ford management decided to install, at the receiving docks, computer terminals capable of reading packing slips. The technology for reading the slips is a more sophisticated cousin of the method used in bar code scanners at supermarkets. Now when an order arrives, a Ford employee has the computer read the packing slip, the computer confirms that the price on the slip is the one Ford agreed to pay, and then as soon as everything clears in the computer system, the checks go out, written automatically by another machine.

The improvement thrilled Ford's suppliers. Reducing the steps in the process and automating operations have resulted in quicker, more accurate payments. Ford management is pleased, too. It reduced the number of employees tied up with approving and issuing of checks by 75 percent. The success of these changes depended on cooperation

---

*This section was prepared by Bruce D. Sanders, Ph.D. For more information on applying automation, we recommend his book, *Computer Confidence: A Human Approach to Computers*, Springer-Verlag, New York, 1984.

among people from purchasing, manufacturing, finance, information services, personnel, and the union. Before giving their approval, the decision makers in finance wanted to be convinced that the dollar savings from the new procedure exceeded the bank interest lost from paying out cash to the suppliers more quickly. Both finance and information services departments required verification methods to prevent fraud. Union representatives insisted that the loss of job positions be accomplished through attrition and employee-retraining programs. Throughout, it was clear that the supplier payment process could be improved only with trust and collaboration. In any such endeavor, teamwork is essential.

As part of BPI, your PITs develop flowcharts. From these charts, you will be able to spot various operations that might be automated. In deciding where to begin, look for:

- Repetitive operations that will improve if performed more quickly. Computers are best at quickly completing routine tasks. On the other hand, people offer opportunities for variety and a change of pace. Give the repetitive work to the computers, especially if you want fast results.
- Operations that will improve when people who are physically separated communicate more quickly. The right kinds of computer systems allow prospects to give information directly to your organization's marketing department. They enable an employee who discovers a mistake to promptly notify the person responsible.
- Operations for which standardized computer system components are available. Want to hear some horror stories? Then start asking about what happened when someone selected oddball computer hardware and software. To avoid your own horror stories, start out by automating those operations in which you can use up-to-date, off-the-shelf components.

Computer systems can be used to facilitate communications between customers and the company. To improve order entry, Caterpillar Industrial Inc., a manufacturer of lift trucks and tractors, hooked into a telecommunications system. Now customers can instruct their computers to telephone Caterpillar and call up an order form onto the computer terminal. The simple structure of the form enables customers to order hundreds of parts with a minimum of keystrokes. Because the orders are entered directly, both errors and delivery delays are cut significantly.

Because computers have massive storage capacities, records of problems can be captured and analyzed, and the responsible person notified. But a better aim is to catch problems as soon as they develop. Across long distances, computer systems use telecommunications networks, perhaps sending the signals over telephone lines or bouncing the signals off a satellite in stationary orbit. Within a building or group of buildings, computer systems communicate over local area networks.

Automobile manufacturer Honda uses a computer network to manage inventories. When a parts clerk at a dealership places an order on a computer terminal, the system immediately polls 10 regional warehouses. That speeds up delivery of parts. Automating the process in this way also allows Honda to keep fewer parts on hand because what's not in one warehouse is likely to be in another. A similar network at Digital Equipment Corporation slashed inventory carrying costs in half over a 5-year period.

Computer networks, when combined with a sophisticated database management system, make it easier for people to retrieve information without needing to know how and where the information is stored. Answers to your questions may be on a hard disk hooked to a personal computer, inside the memory of a large mainframe, or somewhere else entirely. You just request the information. The system figures out where it is and how to send it to your terminal or printer. You get what you need without delay.

Combine telecommunications, local area networks, and database management systems. Then mix in one more type of technology that keeps people in touch—portable computers—and you have a very powerful combination.

Using portable computers and telephone lines, Hewlett-Packard sales representatives can quickly pull out needed information from the company's large mainframe machine. This technology has reduced meeting time more than 5 percent, providing extra time to contact customers. The system also enables sales representatives to retrieve account histories and order status while at the customer's office. The tone changes from one of persuading a prospect to one of collaborating with a client; this increases customer satisfaction and builds sales.

Automation Aims

The only constant is change. This is certainly true with automation. For many products, computer manufacturers plan in 3-year marketing life cycles. Within 3 years of release, the hardware will be obsolete. Software vendors must face consumers who have purchased a program only to see an updated version released within weeks. These suppliers now promise current purchasers new versions for a nominal cost. Commercial data banks—collections of information provided by outside suppliers—grow in scope and ease of use with each passing month. For the consumer of automation, what is not cost effective this year might be quite economical next year.

Waiting for a stable state in office automation would be to wait forever. As with other BPI decisions, automation consists of a series of steps from where you are now to where you want to be, acknowledging that changes will inevitably occur as you go.

Implementation of Automation

Successful automation builds on pilot projects with involved participants. The projects are selected because they show signs of automation opportunity. The participants should be selected with equal care. Look for those who have the aptitudes, abilities, and attitudes to use automation productively. In addition, spot the people willing to give promising change a chance. Organizational psychologists refer to such workers as *early innovators*. Avoid individuals who would give any kind of change a chance. They lack the critical judgment you'll need during project implementation. Perhaps even more important, these employees often lack credibility with their coworkers.

Once you have selected the project and the people, take the time to provide adequate training in the automated operations. Then once you begin, run in parallel. That is, keep the old system going while you try out the new automation. From a systems management perspective, this allows you to use the output from the old system to check the output from the new system. If they don't match, something needs fixing. From a human management perspective, running parallel reduces the stress of the changeover for the employees. They feel that they can fall back on the familiar system if necessary.

But above all, keep a firm focus on the goal of BPI. That's what American President Lines (APL) does. APL's goal is to build business by assuring customers they'll receive predictable, timely delivery of their shipments. To support this goal, the marketing staff uses a network of desktop computers which identifies APL's most profitable customers in the most lucrative markets. Then APL's valuable resources—ships, trucks, rail cars, and personnel—are devoted to assuring the best of services to those premier business teammates.

For example, a customer who wants to know the location of a shipment can telephone an APL computer connected to a network of space satellite and land-based communications links. Within 10 seconds, a computer-generated voice reassures with a precisely accurate answer.

APL is among the few international shipping companies that are consistently profitable. U.S. Lines Inc., formerly comparable to APL in many respects, has left the business. In 1986, APL's ocean cargo volume was up 28 percent in a market that grew only 17 percent. In 1988, APL announced the purchase of five new ships as part of a $500 million company expansion.

THE RIGHT MARRIAGE OF PEOPLE AND PROCESS

When activities are combined, simplified, standardized, or automated, the related skills required also will change. Typically, these changes will necessitate an increase in the number of entry-level professionals and reduce the number of senior professionals. The organization should plan for the changes in mix of professionals, or senior-level employees will end up doing routine and dull work. This adds to the cost of the process, not to mention the frustration involved.

One good technique for assessing the required skill level works as follows:

- Identify the tasks to be performed in the new streamlined process, and estimate the effort required for each task per month based on current work load.
- Create job levels for the purposes of this analysis. Three to four levels (such as senior-level professionals with more than 10 years' experience, mid-level professionals with 5 to 10 years' experience, entry-level professionals with less than 5 years' experience, and clerical employees).
- Evaluate the minimum skill level required to perform each task by asking: What is the lowest job level that can adequately perform the task?
- We now have information on tasks, effort, and skill level required. Summarize the effort by skill level.
- Compare this with the job levels of people actually supporting the process (see Figure 6.4). The mismatch identifies how future adjustments to the organization's skills should take place.

REAPING THE BENEFITS OF STREAMLINING

Once the process has been analyzed and streamlined, there will be several benefits in improved efficiency, effectiveness, and adaptability:

- The customers will get what they want when they want it.
- Cycle time of the process will be reduced.
- Space requirements will be reduced.
- The number of steps and approvals will be reduced.
- Noncritical output will be reduced.
- Cost of the process will be reduced.
- Cost of management will be reduced. (When processes are streamlined, we have, in effect, reduced many of the activities management performs to support the process.)

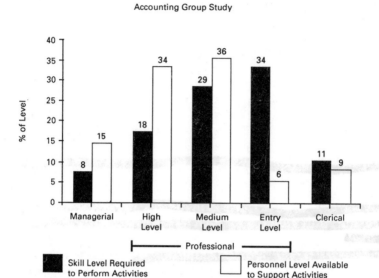

Figure 6.4 Skill-level analysis.

It therefore stands to reason that we should streamline both the process and the management structure that supports it. This can be done by reevaluating every managerial position to determine whether activities are needed under the new process. The current span of control can also be compared with a maximum span of control. Several models are available that can recommend a maximum span based on factors such as job complexity, criticality, employee experience, and geographic dispersion. This analysis would typically point out a need for reducing the number of managers and organizational levels.

SUMMARY

Business processes never remain static. They either improve or they deteriorate. Streamlining is one way to improve the performance of your organization and achieve the goals of your team. You should have two primary goals:

• Develop practical and effective principles to follow in improving work methods
• Develop an organized approach to improvement, from identification of opportunities to the implementation of the desired change

Improvement does not mean increased work load. It does mean eliminating meaningless activity in the jobs, as well as roadblocks to good performance, and some of the frustrations. It means completing work more easily, safely, and efficiently, with fewer errors. It means understanding more about the process and its results.

The cumulative effects of applying the principles of streamlining are amazing. Typical effects are:

- Eliminating bureaucracy removes a major roadblock to performance and high morale, and it reduces costs.
- Reducing NVA activity decreases the number of useless tasks, which makes work more meaningful and productive. It also reduces costs.
- Simplifying the process and subprocess facilitates the work and the understanding; also, processing times are reduced and customers are better served. Again, costs are reduced.
- Error proofing the processes and activities means reduced error rates, fewer crises of the moment, and fewer demands on management and support personnel. Again, costs are reduced.

Don't underestimate the resistance to change the PIT will encounter, if top management has not set the stage. I close this chapter with a comment made by one of the technical reviewers of this book.

> Rational explanations are not likely to overcome resistance in an entrenched, self-protective bureaucracy. A consultant can utilize the author's suggestions to convince an executive officer on the needs for streamlining and upgrading, but the PIT leaders will need extraordinary abilities to implement such changes as they run into inevitable apathy and fears.

This is good advice from an experienced individual who obviously works in an environment that has not totally embraced change as a way of life. There are still far too many of these types of organizations and people in the world—people who look out for themselves and do not look at the total picture. These people need to be identified so that management can help them change. In the rare case when an individual cannot adapt, management needs to assign the individual to a place where he or she cannot slow down or stop progress. Change is always difficult. The executive improvement team (EIT) must set the stage for the change process. If the PIT runs into resistance or lack of cooperation, the situation should be immediately corrected by the EIT.

7

Measurements, Feedback, and Action (Load, Aim, and Fire)

INTRODUCTION

There is no doubt about it. Case after case demonstrate that the lack of good white-collar measurements is a major obstacle to improved business processes. Every experienced manager knows that providing performance feedback to every employee is an essential part of any improvement process. They realize that if you cannot measure it, you cannot control it. And if you cannot control it, you cannot manage it. It's as simple as that. We have already learned how to develop customer measurements, but now we need to probe deep into the process so that we can ensure that the end output will be good and that all parts of the process are improving. In-process measurements provide windows through which the process can be observed and monitored. These windows must be dependable and provide a continuous view of the process. Without dependable measurement, intelligent decisions cannot be made. This chapter looks at how measurement systems work, and how they are used in business processes. We will discuss the 11 W's:

1. Why you should measure
2. Where you should measure

3. What you should measure
4. When you should measure
5. Who should be measured
6. Who should do the measuring
7. Who should provide feedback
8. Who should audit
9. Who should set business targets (standards)
10. Who should set challenge targets
11. What should be done to solve problems

Measurements are critical to:

- Understanding what is occurring
- Evaluating the need for change
- Evaluating the impact of change
- Ensuring that gains made are not lost
- Correcting out-of-control conditions
- Setting priorities
- Deciding when to increase responsibilities
- Determining when to provide additional training
- Planning to meet new customer expectations
- Providing realistic schedules

BENEFITS OF MEASUREMENT

Why is it that an employee who complains about how hard he or she has worked all day will then go home and play three sets of tennis (where he or she will expend twice as much energy as during 8 hours at work), and love doing it? It is because he or she feels a sense of accomplishment when the measurement system gives direct feedback. Why don't people just run back and forth across their driveway, bouncing a ball with their $250 tennis racket? They would get the same amount of exercise and the same experience of coordinating their movement to make contact with the ball. The fact is that they don't get the same sense of accomplishment out of bouncing a ball in the driveway because there are no set rules, no one is involved to see how they are performing, and there is no winning or losing.

Dr. Charles Coonradt, a management consultant, asks, "Why is it that your employees will pay to work harder than you can pay them to work?" Certainly, one of the reasons is that a well-defined measurement system exists to provide immediate and meaningful feedback. The thrill of bowling is not in throwing the ball; it's in knowing how many pins you knocked down.

On the whole, management does a poor job of providing business process measurements. A 1988 survey by the American Productivity and Quality Center found that only 38.7 percent of employees thought that there were good, fair performance measures where they worked. People want to be measured. They need to be measured. The only people who don't like to be measured are the poor performers. In fact, if management fails to establish appropriate systems, the good performers will develop ways to measure themselves to show you how well they are doing. Frequently, however, these measurements are not those that are really important to the business. Management must work with the employees to develop measurements that have meaning for both the employee and the company.

In sports, the rules are well defined. No one can bowl more than 300 points in a game. The same is true of business processes. The procedures (rules) limit how good we can be. The PIT's role is to develop a set of procedures and processes that allows employees to do their best. The employee's responsibility is to maximize his or her performance so that the output is as close to the process design limits as possible.

Why measure? Without it, you take away an individual's feeling of accomplishment, and you never know whom to fire or whom to promote. Measurements, and a good reward system, stimulate the individual and the team to make the additional effort that sets the organization apart from the ordinary.

Lord Kelvin said, "When you can measure what you are speaking about and express it in numbers, you know something about it, and when you cannot measure it, when you cannot express it in numbers, your knowledge is of a meager and unsatisfactory kind. It may be the beginning of knowledge, but you have scarcely in your thoughts advanced to the stage of science."

As important as measurement is, by itself it is worthless. Unless an effective feedback system exists, measurement is a waste of time, effort, and money. Specific feedback enables an individual to react to the data and correct any problems.

For example, when an IBM plant in Havant, England, began to provide each assembly operator with final unit test feedback on the units he or she produced, a major decrease in defects resulted. Figure 7.1 shows two sets of bar graphs that plot the percentage of defective machines by a responsible operator. The letters on the bars are the initials of the responsible operators.

Such dramatic results are not limited to manufacturing. Another corporation that used weekly time record logs to pay its employees and to determine how much overtime and overtime meal allowances to pay found that, initially, over 15 percent of these documents were either not turned in on time or turned in with errors, causing a great deal of additional handling. A simple chart was provided to each manager docu-

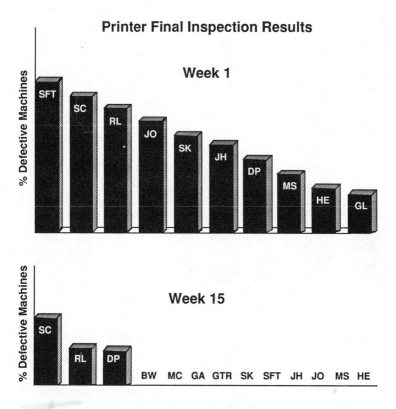

Note: In week 15, 97 percent of the machines were defect-free.

Figure 7.1   Typical results of individual feedback.

menting the percentage of defective time record logs for each department within each function (see Figure 7.2). As a result of this data
feedback and the associated corrective action, the time recording errors
dropped from 15 percent to 1 percent in less than 6 months.

UNDERSTANDING MEASUREMENTS

Why You Should Measure

Measurement is fundamental to our way of life. We measure everything. We measure our lives in seconds, minutes, hours, days, and
months. We measure how far we travel in miles, the food we buy in
ounces or grams, the milk we drink in quarts. Measurements are such
an integral part of our lives that we couldn't manage without them.

When we were babies, the doctor measured our height and weight

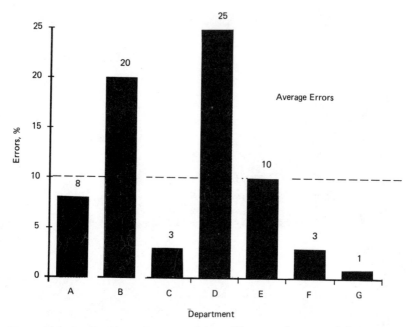

Figure 7.2 Feedback on time record logs. (January time record log errors. Function: manufacturing.)

to be sure that we were healthy. When we started school, our teachers measured us to understand our weaknesses and help us progress. Probably the single biggest thing that enabled Japan to progress so rapidly is the way it used measurements and competition between individuals to get children to excel in their education system. United Nations studies consistently rate Japanese students no. 1, which is because of the fierce competition for grades. As we leave school and go to work, measurement continues. At work, our worth is measured by our salary, and how well we perform.

Where You Should Measure

The major problem with most business processes is that performance is measured only at the end. In most cases, this provides little relative feedback about individual activities within the process or, when it does, it is too late. The process improvement team (PIT) should establish measurement points close to each activity so that the people performing each separate activity receive direct, immediate, and relevant feedback. Consider, for example, how difficult it would be to manage your long-distance telephone bill if all calls in the organization were charged to the same account number.

When You Should Measure

Measure as soon as the activity has been completed. Don't run your business like someone who does not record the amount of the checks he or she writes but just waits until the bank statement arrives to determine what the balance is. Delaying measurement only allows additional errors to be made.

What You Should Measure

In Chapter 3, we discussed the importance of measuring the efficiency, effectiveness, and adaptability of the total process. The same is true of each operation. For example, typical measurements for a sales process might be:

- Sales versus objective
- Percentage of quotation errors
- Number of unpaid invoices over 20 days old
- Percentage of lost business
- Percentage of unanswered phone calls in 4 hours
- Percentage of letters retyped
- Dollars not paid versus dollars billed
- Time between order receipt and order entry
- Percentage of rush orders

For additional information, see Dorsey J. Talley's book *Total Quality Management—Cost and Performance Measures.*

Who Should Be Measured

Management is responsible for providing sound measurement systems, and appropriate feedback, to help all do their jobs better. Management signals what is important by measuring the results. Many people figure that if it isn't measured, why do it? Consequently, every job can and should be measured. Although, theoretically, each task should be measured and reported to the individual performing the task, this is not always practical. The best way to start the process is to examine each activity in the flowchart and identify those that significantly impact total process efficiency and effectiveness. Then establish measurements for these critical activities.

Reviewing the internal customer satisfaction level will identify a second set of measurement priorities, focusing on activities that are not meeting internal customer expectations. Give third priority to activities that require significant resources and those that provide an individual's only job performance feedback. You can measure performance in terms of effectiveness and efficiency and express it in physical terms (e.g., time to perform a task, cycle time) or in dollars, allowing a number of resources to be combined (e.g., value-added costs, labor costs).

When developing your performance measurements, get help. Ask your customers what is important to them, ask your employees what is meaningful to them, ask your suppliers what is fair to them, and get their agreement with and support of the family of measurements you decide on.

Who Should Do the Measuring

The best person to do the measuring is the person performing the activity. There is immediate feedback, and he or she should have the best understanding of the job. When the self-inspection error rate is too high, let the people performing the activity check each other's work. They will learn this way, which is a good way to exchange ideas and to begin standardizing. As a last resort, have someone who is not part of the activity check the output. Whoever does the measuring should be well trained and use documented criteria to evaluate the output.

It takes almost as much time to evaluate an output as it does to perform many of the activities. All evaluations are expensive. It takes time to proofread a letter, to check an expense account, to do a design review. Even self-inspection must be planned for. Consider how long it would take a second party to review the output, and allow the individual performing the activity 50 percent to 60 percent of that time to do the self-review.

Experience has shown that while the time to do a job is understood and, in most cases, considered in the planning, little or no consideration usually is given to the time required to measure the results and make appropriate corrections. The total cycle time must be considered and planned for if you don't want to have the quality of the total process compromised.

Who Should Provide Feedback

This is the supplier-customer partnership theory we have talked about. Each output receiver should provide positive and negative feedback, and constructive criticism, to the person or people providing the product or service.

Who Should Audit

We audit manufacturing personnel to ensure that they do their jobs right. We audit finance to ensure that we are not being fleeced. Doesn't it make sense, then, to audit our business practices using an independent party? Or, at the very least, shouldn't the management in each area conduct regular, formal, documented audits using written audit procedures? The results of these audits should be reported to management, and to the employees, along with appropriate corrective action.

Unfortunately, few companies possess the disciplined management structure, and the level of trust, required for an effective self-assessment audit system. If the self-assessment system is inappropriate, the systems assurance group should conduct an independent audit.*

Who Should Set Business Targets (Standards)

How effective should we be? How efficient should we be? Who should set the standards within the process? How do we know if we are doing well? How do you know that your manager isn't getting ready to fire you for doing a poor job? Who makes the rules?

Consider business targets (standards) and what they mean. A standard sets the minimum acceptable performance for an individual performing an activity. It is the result that the present process will produce with a person who has been trained to do the job, has the necessary tools, and has the ability to do the job. It is not the present performance level. In most cases, the standard will be lower than the present performance level, if you have an experienced person doing the job. When the standard is not met, there is something wrong, and you should take action. Remember that the process design sets the maximum performance level. Your employees cannot perform better than the process allows.

Let's look at the two key measurement types—efficiency and effectiveness—and how to set standards for each of them. First, let's address effectiveness (quality). The person or people who receive the output (the internal and external customers) should set the effectiveness standard. You should meet with these customers and determine exactly what they need. Then design a process that a below-average person can use and still meet the standard. Remember, 50 percent of your employees are below average, so design your processes to operate using below-average employees. Efficiency (productivity), in contrast, is not customer driven. It is controlled by the process. All processes have an inherent efficiency—the least amount of resources required to provide the output when everything goes right. But things don't always go right. Waste and inefficiency are built into the process.

Consequently, the scientific approach of task-time analysis usually concedes to a much less exacting process of having the employee and manager review past performance data to establish an efficiency stan-

---

The Improvement Process, chaps. 5 and 10.

dard. The manager then must decide if the value-added content is worth the price. If it is not a good value, either the process is redesigned or the manager decides not to perform the activity. Such standards often are based on the present employee's skill in using the process and may be unrealistic when a less-skilled employee performs the activity. Determine what resources are justified to complete the specific activity, based on the value of the output. Then design the process to provide the desired performance level a high percentage of the time with below-average employees.

Who Should Set Challenge Targets

Don't confuse business targets (standards) and challenge targets. They are very different. Business targets are set by the customers, since they reflect their expectations, or by management, since they define minimum acceptable performance. A challenge target, on the other hand, is an objective set by the team or the individual performing the activity. A challenge target is always more stringent than a business target and supports the concept of continuous improvement. It is the means for providing customers with surprisingly good quality. Failing to meet a challenge target should not impact the business plan. Your objective always should be to be better than your customer expectations, to be better than your business plan, and to accomplish more with fewer resources.

Once a business target is met, the team, or the individual, should set challenge targets to stretch them to meet new, higher levels of performance.

MEASUREMENTS ARE KEY TO IMPROVING

Dr. Coonradt believes that we should make business more like a game. He points out that the same workers who require special clothing to work in the cold storage section of a plant for only 20 minutes, with a 10-minute break to get warmed up, will spend an entire day ice fishing on a frozen lake. When the air-conditioning breaks down, and the temperature reaches 85 degrees, the office is closed. However, on the way home, employees will stop and play a round of golf in the same heat—and pay to do it.

Why is this? Sports succeed in exciting people because they have rules, measurements, and rewards. Let's apply these principles to business:

Rules. Every sport has rules that govern the game. People know them and are penalized when they do not live up to them (e.g., a 15-yard penalty for clipping another player in football). Business, too, has its rules. They are the procedures and job descriptions. Business also has referees—the quality system auditors. When you do not play by the rules, you should be penalized.

Measurements.   How much fun would it be to play tennis and not know whether the ball landed in the other side of the court? We need to know how well we are doing, and it needs to be personal. How popular do you think golf would be if the only feedback you received is that, of the 200 people who played golf last Sunday, the average score was 93, and you weren't told what your score was. Yes, measurements are critical to maintain interest in an activity, particularly when you want to improve. Measure both individual and team performance. A baseball team, not an individual, wins or loses the game. Apply team measurements to small groups (20 maximum).

Rewards.   Sports enthusiasts get their rewards from trying to improve their games. The professionals see big dollars rolling in, and the amateurs win trophies, but all are tied into a measurement system. In business, we should all be pros, or we should not have our jobs. Sure, it is nice to get the trophies, plaques, and dinners when we excel, but we also should receive financial awards. Our pay should be tied directly to our personal measurement system.

Measurement is important for improvement for several reasons:

- It focuses attention on factors contributing to achieving the organization's mission.
- It shows how effectively we use our resources.
- It assists in setting goals and monitoring trends.
- It provides the input for analyzing root causes and sources of errors.
- It identifies opportunities for ongoing improvement.
- It gives employees a sense of accomplishment.
- It provides a means of knowing whether you are winning or losing.
- It helps monitor progress.

Business process organizations tend to think, and maybe even believe, that they cannot be measured. Consequently, in the past we measured only products and ignored business processes. This happened for several reasons:

- In the 1950s and 1960s, product cost (material and labor) constituted a significant portion of total cost. Hence, management was preoccupied with measuring and managing it. Business process cost was a smaller element of cost and therefore was ignored.
- Traditional measures of input and output could not be applied easily to business processes.

- White-collar workers believed that their work was varied and unique and could not be measured.

Measurement trends are changing:

- From product measurement to process and service measurement
- From managing profits to managing assets
- From meeting targets to continuous improvement
- From quantity measurements to measurements focusing on effectiveness, efficiency, and adaptability
- From measurements based on engineering and business specifications to measurements based on internal and external customer expectations
- From a focus on the individual to a focus on the process (it was assumed that individuals could control all results; now we accept that it is the process that should be measured)
- From a top-down process dictator approach to a team approach to developing measurements and managing performance

Obviously, all business processes can, and should, be measured and managed in the same way manufacturing processes are.

TYPES OF MEASUREMENT DATA

As you establish your measurement system, the PIT should work with two types of data: attributes data and variables data.

Attributes Data

These kinds of data are counted, not measured. Generally, attributes data require large sample sizes to be useful. They are collected when all you need to know is yes or no, go or no-go, or accept or reject. Examples of attributes data include:

- Did an employee arrive at work on time?
- Was the letter typed with no errors?
- Is a department below budget?
- Did the meeting start on schedule?
- Was the report turned in on schedule?
- Was the phone answered by the second ring?

Variables Data

Variables data measurements provide a more detailed history of your business process. This involves collecting numeric values that quantify a measurement and therefore requires smaller samples. Examples of variables data include:

- Number of times a phone rings before it is answered
- Cost of overnight mail
- Number of hours to process an engineering change request
- Dollar value of stock
- Number of days employees call in sick per year
- Number of days it takes to solve a problem

Occasionally, it may seem difficult to establish meaningful measurements (e.g., How good was a presentation? How well did a document reproduce? How clean is the office?). In many situations, human judgment enters the picture, and you can compare relative values. For example, you can judge print quality by comparing a number of pictures. In some cases, there is no other way but to ask your customers for their opinions on some of the softer measurements. After all, their opinions are the real measure of quality.

Understanding Attributes and Variables Measurements

You can measure most things using attributes or variables data. It usually takes a bit longer to collect variables data, but often the picture provided is much more revealing. Figure 7.3 lists typical examples of measurements made using both attributes and variables data. You get a lot more information about the output when you collect variables data, and it is a lot more useful.

Attributes measurements	Variables measurements
Did the employee arrive to work on time?	How many minutes late was the employee?
Was the letter typed with no errors?	How many errors were there per 100 words recorded?
Is the department below budget?	How big are the variations between actual expenditures and budget?
Did the meeting start on schedule?	How many minutes late did the meeting start?
Was the report turned in on schedule?	How many hours early or late was the report turned in?
Was the phone answered on or before the third ring?	How many times did the phone ring before it was answered?

Figure 7.3 Examples of measurements using attributes and variables data.

Let's compare two reports in which 100 telephone calls were monitored using both variables and attributes data. The target was to answer all telephone calls before the third ring. The attributes data simply would report that 45 percent of the calls were answered on or before the third ring. The variables report would take the form of Figure 7.4. The two pieces of data tell different stories. For example, the variables data point out that the company is losing sales because the telephone is not answered 27 percent of the time. Figure 7.5 presents the variables analysis of the unanswered calls.

Based on this analysis, the customer requirement seems to be answering the phone within five rings. Their expectation may be three, but their requirement is that the phone be answered before it rings six times. Based on this data, the standard should be to answer all telephone calls before the sixth ring; and the challenge target should be to answer them before the fourth ring.

Effectiveness Measurements

Effectiveness measurements are the results obtained from the resources expended. They are often connected directly to internal and/or external customers and indicate how well the output from an activity or

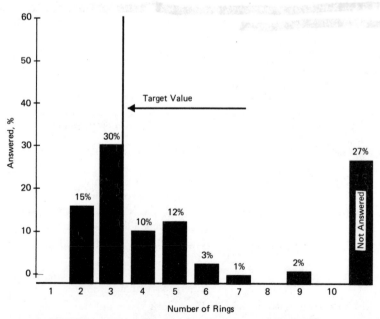

Figure 7.4 Number of rings before the telephone is answered.

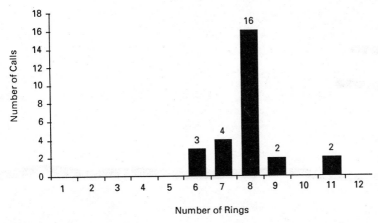

Figure 7.5   Number of rings before caller hangs up.

group of activities satisfies customer expectations. An effective process consistently produces products and services that meet or exceed customer needs and expectations with low process variability.

To establish effectiveness measurements, consider what the customer wants:

- Accuracy
- Timeliness
- Dependability
- Responsive people servicing them:
  Caring
  Polite
  Knowledgeable
  Empowered

Let's take product engineering as an example:

Expectation	Measurement
Accuracy	Number of engineering changes per drawing
	Number of problems found in preship evaluation
	Number of customer complaints related to the design
Timeliness	Time to solve a problem
	Variation of the actual first customer ship date of product or service from marketing's original definition of the customer window

Expectation	Measurement
Dependability	Percentage of missed target dates
	Percentage of meetings that start on time
Responsiveness	Percentage of time with customer
	Percentage of time on the manufacturing floor
	Percentage of inquiries correctly answered on the spot
	Number of days to correct a problem

Production control is another area where effectiveness is incorrectly measured. At the weekly production meeting, it is reported that 99.8 percent of the parts are on the manufacturing floor for the coming week. Two parts are missing, and production control rushes in like the cavalry to save the day by expediting the parts over the weekend. They are viewed as heroes, when in truth they are the villains—because they caused the problem in the first place. As long as they are treated as heroes for solving problems they create, they are going to keep creating them.

Their measurements should be:

- Percentage of reports with errors
- Percentage of parts delivered late
- Percentage of weeks that all parts are on the production floor on time
- How fast they can respond to a schedule change

Other types of effectiveness measurements that should now be considered are:

- Return on investment
- Reliability
- Number of new customers
- Morale

Efficiency Measurements

Efficiency measurements reflect the resources that an activity or group of activities consumes to provide an output meeting internal and/or external customer expectations. An efficient process is one where resources are minimized, and waste eliminated.

Typical efficiency measures for an expense reimbursement process are:

Throughput	Volume of transactions per employee processing them
	Value-added time spent as a percentage of total time
People utilization	Labor hours per 100 transactions
Computer utilization	Machine minutes per 100 transactions
Cost reduction	Cost to process a transaction
	Time to fill out the form
	Number of documents used per transaction

Adaptability Measurements

Adaptability measurements reflect how well the process and people react to special customer requests or the changing environment. Because of the long cycle time required to adapt to environmental changes, the measurement system should address the ability of the process to react quickly to the special internal and/or external requests.

Adaptability is hard to measure but must not be overlooked. Customers remember how you went out of your way to help them when they needed it. It's how a company reacts to special needs that sets apart the outstanding supplier from the merely good.

Far too often, we fail our customers by saying, "that's company policy" or "that's the procedure." I will never forget a salesperson I met when I was abroad. I was in the midst of selecting a gift to bring home to my wife when I was told, "Sorry, that's the lunch bell. I have to go to lunch at 12:00." Off she went, leaving me, money in hand, to find another salesperson on duty. Such things happen far too often. Airlines give up your seat because you didn't arrive 10 minutes before the flight's scheduled takeoff time, even though a snowstorm held up your connecting flight for 45 minutes, and the flight on which you were booked is scheduled to leave 1 hour late. Stores will give you only store credit for a gift you received that didn't fit, even though you live 1000 miles away. Repair stores won't repair your television set because you didn't bring your warranty with you, although they know the model has been available for only 1 year—and the set comes with a 2-year warranty.

Far too often, we use procedures as an excuse to save us from making that little extra effort that would result in a happy, satisfied customer. At the same time, employees who deal with internal and external customers are not empowered to use their own common sense. They are not allowed to treat people the way they, themselves, would like to be treated. The authority to deviate from the established processes is viewed as the exclusive right of the management, and this is wrong. We must empower and trust our employees to make the proper decisions required to satisfy special customer requests.

Employees should take it upon themselves to deviate from the process when they realize that they have a customer who is dissatisfied. Typical examples would be reducing the price of a replacement article, when the customer had problems with the one he or she returned, or not charging for a meal because the potatoes were cold. Too often we worry about pennies, and we lose dollars. Would a supermarket argue about whether or not a loaf of bread is fresh if it stopped to think about the thousands of dollars it would lose if the customer stopped shopping at its store?

This is what it costs different industries to lose a customer (based on 1988 costs).

Automobile	$140,000 in a lifetime
Appliance	$160 per year
Supermarket	$4400 per year
Hospital	$60,000 in a lifetime

The very best way to keep a dissatisfied customer is to have empowered employees that can take action to do something special to make up for the discomfort the customer experienced.

Adaptability measurements identify how often requests are made to deviate from the prescribed process, the percentage of time customer requests are granted, and at what level the deviation requests are granted. Typical adaptability measures are:

- Number of special requests per month
- Percentage of special requests granted
- Percentage of special requests granted by direct contact employees
- Percentage of special requests self-generated by direct contact employees

Closely monitoring special requests is a good way to determine when to change the basic process.

PROCESS MEASUREMENT SYSTEMS

Manufacturing processes typically institute a quality control system to ensure that the processes and the products meet the defined requirements. Similarly, once business processes initially have been streamlined, they should be maintained so that they do not revert to their original form. They also should be improved further so that your organization gets and stays ahead of its competition. You can do this by developing a measurement and feedback system for the business process. The steps involved in developing such a system are:

- *Establishing effectiveness and efficiency requirements for the process.* These requirements indicate the performance level of the process and typically are based on internal and external customer and business requirements. In Chapter 3, we defined the requirements for the total business process prior to analyzing and streamlining the process. The PIT should review those requirements, based on additional knowledge of the process, to ensure that they are still valid.

- *Establishing an in-process measurement and feedback system.* Once process quality requirements are set, we need to focus on the subprocesses, activities, and departments contributing to the overall process. Only if these lower-level processes work efficiently and effectively will the process quality requirements be met. Here again, the flowchart will help you to identify customers internal to the process. Look for places where output changes hands from one person to another or one department to another. These are all points at which you could establish effectiveness measurements. Of these lower-level processes, only some are critical and have a major impact on the overall quality. At a minimum, establish a measurement system supporting process performance for those critical subprocesses or activities. The ultimate goal for the PIT is to establish measurement and feedback systems for every activity within the process.

There has been a lot of talk about the internal and external supplier-customer relationship. The supplier should supply high-quality output that meets or exceeds customer expectations on schedule, at a fair price. But what is the customer obligation in this partnership? A partnership indicates that both parties gain from the relationship. It should not be a one-way street. The customer's role is to provide feedback on how well the supplier is performing. External customers provide this feedback in the form of money and repeat purchases. Internal customers have the same obligation to provide feedback to the internal supplier. In most companies, the internal customers do not pay for the service, nor do they have the option of going to another supplier. Because the internal supplier has no other way of obtaining performance data, the internal feedback system is extremely important. On the whole, internal customers have done a poor job of living up to their side of the partnership. The business process measurement and feedback system recognizes the obligation of both the supplier and the customer.

Now look at the process to define efficiency measurement points. First establish efficiency measurements for the parts of the process that use large amounts of resources (material, equipment, employee hours). Then look for subprocesses or activities that cause long total cycle time, and put in place another set of efficiency measurements. All but very minor activities should have efficiency measurements developed for them.

■ *Setting business targets.* Once measurement systems are defined, the question is: What is acceptable performance? There is only one person who should and could answer that question—the person who receives the output. The best way to establish internal effectiveness targets is to have the individual performing the activity meet with his or her customer(s) and ask them to set the minimum acceptable effectiveness target. Efficiency targets should be set jointly between the employees doing the job and the manager of the area.

We have talked a lot about continuous improvement. If continuous improvement is the organization's goal, why not do away with targets and just evaluate progress based on the direction of the trend line? Well, that is a good point. It certainly is the mind set we are looking for. But if your charts don't have targets, the only goal you will have is to be perfect. If your only objective is never to make another error, each time you do make an error, you have failed, and that soon becomes unmotivating.

It's a lot like climbing a mountain. The goal of mountain climbers is to reach the top, and they develop a plan to lead them there. On the way, however, they also must set targets for getting around an over-hang, or up a chimney, to a point on the mountain where they can rest before taking on the next challenge.

Targets are necessary because they provide a sense of accomplishment, a point for recognition. They are the stepping-stones of success that lead up the road to perfection. They turn what could be a discouraging trip into one that is fun and rewarding. Targets provide a challenge to the team, and the individual, and encourage continuous improvement. In fact, targets often set the angle of the continuous improvement trend line. Challenging targets set by the people responsible for obtaining them frequently result in very rapid improvement. When management sets the targets, progress will be much slower since, frequently, these targets are much less aggressive than the targets the employees set for themselves. In addition, when management sets the targets, the employees have less invested than when they, or the customers, set them. Another, and even more important, reason for targets is to ensure you are meeting your customers' needs and expectations. Such customer requirements set the minimum performance standard for the organization.

Since the PIT includes members of the entire process, who have studied the process and met with its customers, it should define the measurements for the entire process. Then, at lower levels of the subprocesses, activities, and departments, the PIT should assist the employees involved in each of those areas in developing measurements that are consistent with the overall process measurements and that meet the expectations of their internal customer(s) and management.

Consider the way the measurements will be taken:

- Can you use self-inspection? This is best, because the feedback is instantaneous.
- Will all employees get feedback about their work?
- Will the feedback be timely?
- Are you evaluating the process or output from an activity?

Insist that any assessment or inspection of another's work result in a feedback report summarizing the results of the assessment over a period of time.

THE HUMAN PROCESS

An integral part of every process consists of the individuals involved in the process. In truth, all humans have a number of very complex systems that control processes within themselves that must be considered when you start to refine your business processes. All individuals are very complex processing systems that produce output based on receiving the appropriate inputs and stimulation. Each output from the human process is directed at a receiver (customer). The receiver of the output transmits messages back to the individual, either directly or indirectly, that impact the individual's performance and behaviors. The human will tend to stop doing things that produce negative stimulation and will do things that lead to positive stimulation.

As in any complex process, the human process must have some critical ingredients in place before it can perform correctly. They are:

- Training to perform the task
- An understanding of the desired output and the criteria to measure it
- Physical and mental capabilities to perform the task
- The incentive to perform the task
- The time to perform the task correctly and the necessary tools
- An incentive to perform the task

Once the human process starts feedback, stimulation is essential to keep it going. If the individual is rewarded for doing the wrong things, or punished for doing the right things, the process quickly degrades. If there is no feedback stimulation, or if the feedback systems are late or inadequate, the human processes will not perform in a superior manner. Take away or cut back on the time, tools, or information flow, and the human process quickly becomes inefficient and ineffective. The human process performs as a function of how harmoniously it is connected to the business process.

FEEDBACK SYSTEMS

Feedback systems are very important. It's clear that if you cannot measure an activity, you cannot improve it. But measurement without feedback is worthless because you have expended the appraisal effort but not provided the individual with an opportunity to improve. Measurement is the lock—feedback is the key. Without their interaction, you cannot open the door to improvement.

Feedback Data Quality

Don't underestimate the importance of the quality of data that are fed back to the process. All too often, we make major decisions using erroneous, polluted data. Design the feedback system to provide:

- Meaningful data
- Timely feedback
- Accurate data
- Correct analysis
- Understandable format

To accomplish this, you need to treat the feedback system like any other process. You need to:

- Identify the customer
- Agree on requirements and output specifications
- Eliminate error sources
- Provide for feedback from the user of the data

Establishing Feedback Loops

Consider the following points when establishing the feedback loops:

1. *Relate feedback loops to individuals.*   Try to relate the feedback to the individuals performing the tasks so that they quickly will understand their impact on quality.

2. *Make the feedback an obligation.*   Explain to people why feedback is important and necessary for better performance. Most people are re-

luctant to give feedback, especially negative. In an improvement environment, you must look for errors, not for blame. You must try to learn from your problems and errors. Did you ever go into a restaurant where the soup was too salty or the food was cold and when they asked how the meal was, you said "okay"? Imagine the impact if every dissatisfied customer said, "It's okay." Providing constructive feedback is a consumer's obligation.

3. *Encourage positive and negative feedback.* If you want to support people, give them lots of positive feedback. It reinforces their desire to continue to do a good job. If you want them to improve, provide negative feedback as well. Remind them that most errors are caused by the process, not the person working within the process. Implement a "no-blame" approach. Don't have a feedback system that only reports how bad things are.

4. *Use continuous feedback for continuous improvements.* Organize a systematic and ongoing system to eliminate the situation in which you receive a lot of feedback in the beginning and none later on. Monitor it to ensure that you provide a continuous flow of useful data to the employees.

5. *Avoid the old proverb "no news means good news."* Too many people assume that if they do not receive any feedback, things are okay. No feedback means that you no longer are in touch with your customer. In particular, if you have external customers, it could mean that they have given up on you.

6. *Encourage customer complaints.* Most dissatisfied customers will never tell you that they are dissatisfied. People do not like to give negative feedback. So, you must make it as easy as possible to complain to you. Welcome feedback, and design your system to increase your feedback ratio. Treat each customer complaint as a cherished jewel, a chance to solve a problem, a chance to be better. Don't just handle complaints. Go out of your way to encourage them—then prevent them from recurring.

7. *Give responsibility to take immediate action.* Provide the feedback to the person who can do something about it. Do not send reports just to the manager; send them to the employee as well. Then empower the employee to take immediate action. Require reporting of the corrective action to the person who input the data.

INDEPENDENT DATA AUDIT PROCESS

An essential element of any measurement and feedback system is an independent audit process that ensures compliance to the procedures. It is nonsense to blindly accept data generated without proper checks and balances. Not because people will falsify data—most people won't—but most employees have a strong desire to satisfy management and tell management what it wants to hear.

Often, employees don't understand how data are used or even what data are required. I remember touring one manufacturing plant and spotting a control chart with 6 months of data on it. All the points were within one-quarter sigma of the mean. When I asked about the data's validity, the manager quickly responded, "Mary is one of our best operators. This operation always runs smoothly, just as the graph indicates." However, when I asked Mary to measure five parts for me, all five fell outside of the 2-sigma limit. When I questioned Mary why these parts were so bad, but the results she recorded were so good, she replied, "Well, I have to measure a lot of parts before I can find five that are good enough to record."

REPORTING

The best type of feedback is direct feedback, where employees see the errors they have made and can correct them themselves. Although effective, such feedback often gets lost in the daily routine. Consequently, you should develop a reporting system that provides trend information so that employees can measure their progress. There are many ways to do this:

- The employees can keep the charts themselves (the best way).
- A computer can generate the reports (the fastest way).
- The customers can generate the reports (the high-impact way).

We like to have each major subprocess select at least one measure of effectiveness, and one of efficiency, and post them where the members of the process can see them. We call them *process improvement chart (PIC) indicators.*

Keep the PIC indicators simple and large enough to read from a distance. We recommend using a chart that is at least 24 inches square. Each chart should show at least 6 months of data and include a business target performance level. Once the target is met on a continuous basis, establish a new challenge target (CT). When a chart has met a target for three consecutive measurement periods, reset its targets. Remember, there is no shame associated with not meeting targets; the objective is to constantly improve. Figure 7.6 presents a sample PIC.

We use two types of targets: performance levels that the customer expects (business targets) and tighter CTs. CTs provide the team with interim goals between the customer's minimum acceptable performance level and the ultimate standard of error-free performance. This elimi-

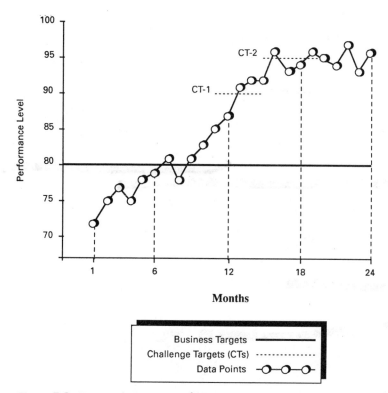

Figure 7.6 Process improvement chart.

nates the tendency of most companies to stop all efforts to improve an activity once the target has been met. It also forces management to look at targets in a new way. Management should expect to meet all business targets 100 percent of the time but should not expect the challenge targets to be met all the time, for as soon as they are met, new challenge targets are set. In fact, they should question managers who are meeting challenge targets to determine why they have not reset the targets.

For best results, PIC indicators should be made in a very professional manner, using pressure-sensitive labels. We recommend designing a standard background customized to each area's needs. A few well-placed PIC indicators can significantly impact performance, increasing awareness and productivity among employees and managers alike.

Once the PIC indicators are in place, it is critical to update them. Few things can be more demoralizing to a team than to have a performance indicator that is out of date. If management does not care enough to keep the PIC indicators up to date, why should the employees care about the excellence of their output? A reliable employee should be responsible for keeping the PIC indicators up to date, and a backup person should be assigned. Review the status of all PICs at each PIT meeting.

Statistical Business Process Controls

Statistics, properly used, allow us to understand how our process is operating. They also prevent us from reacting to nonproblems. People working outside of manufacturing areas often are uncomfortable with statistics, but they should not be. Statistical process control (SPC) is a simple concept that anyone can understand and use. Certainly, the intelligence level of the people working in the manufacturing and assembly process is no better than that of the people who are supporting them or the people who are working in the service industries.

Rather than avoid statistics, we should use this powerful tool to better understand our business processes. SPC helps improve processes by:

1. *Summarizing data.*   Statistics help us to bring our process under control, summarize our data, make accurate assessments, and facilitate decision making.

2. *Providing insight into variability.*   When we refer to a *process*, we mean that activity used to produce a product or service within our work environment. The output from the process always has some variation because of the variation that comes from the people, materials, machines, methods, and environment that make up that specific process.

3. *Clarifying decisions and understanding risks.*   To make good decisions, you need to have good usable data. They must reveal the essential facts and be arranged in a format that is easy to interpret. SPC data often are presented in the form of histograms, bar charts, control charts, and Pareto diagrams.

Process control charts have saved considerable wasted effort in many business processes. At times, process control chart trend analysis has identified potential problems before the internal and/or external customers were aware of them. Analyzing process control charts significantly helps to put your business processes in a problem-prevention operating mode.

There are two types of control charts: attributes data (go or no-go) control charts and variables data control charts. Figure 7.7 lists four control charts that use attributes data. The second type of control chart uses variables data and provides much more detailed information about the process under evaluation. X-bar and R charts are typical variables control charts.

Figure 7.8 presents a typical np attributes control chart used to mea-

Name	Description
U chart	Number of errors per unit
c chart	Number of errors in a sample
P chart	Percentage of outputs with errors when the output is sampled (e.g., 80 percent of the pages in a 100-page sample contained errors)
np chart	Number of outputs with errors when the output was sampled (e.g., 80 pages out of a 100-page sample had errors)

Figure 7.7   Control charts using attributes data.

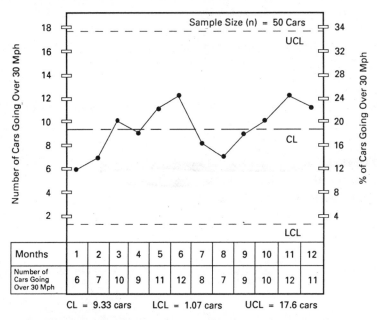

Figure 7.8   An np control chart for cars speeding on company property.

sure how bad the speeding was on company property. You calculate the central line by adding the total number of speeders detected in the checked lots and divide this number by the number of total samples taken to provide an average number of speeders per sample.

The upper control limit (UCL) and lower control limit (LCL) evaluate each point on the graph to determine if the process is under statistical control. Of the points plotted, 99.7 percent should fall between the UCL and the LCL if the process remains unchanged.

Let's consider Figure 7.8, ignoring the center line (CL), the UCL, and the LCL, to see how the data could be misinterpreted when SPC methods are not applied to business processes. (These data were collected and the actions taken before the control limits were applied.)

The safety department reported to management that there was a major problem with speeding on company property. The department recommended hiring a consultant to present a 2-hour class on safe driving to every employee and presented the first 6 months of data to verify the need. Management approved the request, and all 500 employees attended the class in month 7. After the class, speeding errors dropped for 2 months and then rose again.

Management concluded that the class should be repeated every 3 months to keep careful driving an employee priority. However, when a member of one of the PITs plotted the data on a control chart, he proved that the employees' driving habits had not changed. Management agreed to take a different type of action. In month 15, speed bumps were installed, and the average percentage of cars going over 30 miles per hour dropped from 18 percent to 6 percent. The process was permanently improved.

This simple case study vividly demonstrates the need to use SPC in our business processes. There are many places where control charts should be used. They include:

- Number of open doors per month per employee
- Percentage of people reporting late for work
- Time to process a sales order
- Percentage of lost sales
- Number of customer complaints per month
- Percentage of budget expended

Process control charts indicate when the process has changed so that you can correct negative changes before they become a problem. Additionally, they help you to recognize positive changes so that you can capture and maintain these positive shifts in your business processes.

POOR-QUALITY COST

One of the primary objectives of business process improvement (BPI) is to reduce the losses that are caused by poor quality. Poor quality costs your company money. Good quality saves your company money. It's as simple as that. James E. Olson, former president of AT&T, said, "A lot of people say quality costs you too much. It does not. It will cost you less." But many companies today do not measure the cost of poor quality; and if you do not measure it, you cannot control it. Why is it, then, that corporate management does not insist on the same good financial control over poor-quality cost (PQC) as they exercise over the purchase of materials, when often PQC exceeds the total materials budget?

To put it simply, the PQC reporting system is only one of the many tools needed in a comprehensive, companywide improvement process, but it is an important tool since it directs management attention and measures the success of the company's efforts to improve. It also provides management with the necessary tools to ensure that suboptimization does not have a negative effect on the total process.

PQC is defined as all the cost incurred to help the employee do the job right every time, the cost to determine whether the output is acceptable, plus all the cost incurred by the company and the customer because the output did not meet specifications and/or customer expectations. Figure 7.9 lists the elements of PQC.

Why Use PQC?

PQC provides a very useful tool to change the way management and employees think about errors. PQC helps by:

1. Getting management attention. Talking to management in terms of dollars provides them with information that they relate to. It takes quality out of the abstract and makes it a reality that can effectively compete with cost and schedule.

2. Changing the way the employee thinks about errors. There is less impact on an employee's future performance when, as a result of his or her actions, an estimate has to be redone, than when the employee knows that it costs $3000 to redo the estimate. In one case, what is thrown away is a report; in the other case, it's thirty $100 bills that are discarded. Employees need to understand the cost of errors they make.

3. Providing better return on the problem-solving efforts. PQC re-

---

I. Direct PQC
   A. Controllable PQC
      1. Prevention cost
      2. Appraisal cost
   B. Resultant PQC
      1. Internal error cost
      2. External error cost
   C. Equipment PQC
II. Indirect PQC
   A. Customer-incurred cost
   B. Customer-dissatisfaction cost
   C. Loss-of-reputation cost

Figure 7.9   The elements of PQC.

lates problems to dollars so that corrective action can be directed at the solutions that will bring maximum return. James R. Houghton, chairman of Corning Glass Works, has reported, "At Corning, cost of quality is being used to identify opportunities, to help prioritize those opportunities, and to set targets and measure progress. It's a tremendous tool, but we are taking great care to ensure that it is not used as a club."

   4. Providing a means to measure the true impact of corrective action and changes made to improve the process. By focusing on PQC of the total process, suboptimization can be eliminated.

   5. Providing a simple, understandable method of measuring what effect poor quality has on the company and providing an effective way to measure the impact of the improvement process.

   6. Providing a single measurement that brings together efficiency and effectiveness measurements.

Direct PQC

Of the two major PQC categories, direct and indirect, the direct PQCs are the better understood and are traditionally used by management to run the business because the results are less subjective. Direct PQC can be found in the company ledger and can be verified by the company's accountants. These costs include all the costs a company incurs because management is afraid that people will make errors, all the costs incurred because people do make errors, and the costs related to training people so they can do their jobs effectively. Direct PQC encompasses three major expenditures: controllable, resultant, and equipment PQCs.

Controllable PQC

Controllable PQCs are the costs that management has direct control over to ensure that only customer-acceptable products and services are delivered to the customer. Controllable PQCs are further subdivided into two categories: prevention costs and appraisal costs.

   Prevention costs are all the costs expended to prevent errors from being made or, to say it another way, all the costs involved in helping the employee do the job right every time. If you look at this from a financial viewpoint, it is really not a cost. It is an investment in the future, often called a *cost-avoidance investment*. Typical prevention costs are:

   - Developing and implementing a quality data-collecting and data-reporting system
   - Developing the process control plan
   - Quality-related training
   - Job-related training

- Vendor surveys
- Implementing the improvement process
- Design concept reviews

Appraisal costs are the result of evaluating already-completed output and auditing the process to measure conformance to established criteria and procedures. To say it another way, appraisal costs are all the costs expended to determine whether an activity was done right every time. Typical appraisal costs are:

- Outside financial audits
- Approval signatures on a document
- Outside endorsements, such as from Underwriters Laboratories
- Maintenance and calibration of test and inspection equipment
- Review of completed designs
- Review of test and inspection data
- Second-level review of first-level management decision
- Proofreading letters
- Payroll audits

Resultant PQC

Resultant PQCs make up the second direct PQC category. They include all the company-incurred costs that result from errors or, to put it another way, all the money the company spends because all activities were not done right every time. These costs are called *resultant costs* because they are directly related to management decisions made in the controllable PQC category. Resultant costs are divided into two subcategories, internal and external error costs.

Internal error cost is defined as the cost incurred by the company as a result of errors detected before the output is accepted by the external customer. To put it another way, it is the cost the company incurs before a product or service is accepted by the customer because everyone did not do the job right every time. Included are the costs incurred from the time an item is shipped from a supplier until it has been accepted by the final customer. The following are representative examples of internal error costs:

- Retyping letters
- Engineering changes
- Additional costs because bills were paid late
- Costs resulting when additional inventory is required to support poor process yields, potential scrap parts, and rejected lots
- Computer reruns
- Overcooked food

External error cost is incurred by the producer because the external customer is supplied with an unacceptable product or service. It is the cost incurred by the company because the appraisal process did not detect all the errors before the product or service was delivered to the customer. Typical external error costs are related to items such as:

- Cost of customer-rejected services or products
- Product liability suits
- Complaint handling
- Warranty administration
- Repair personnel training
- Product recall or updates in the field
- Overhead costs required to maintain field service centers
- Missed sales

Indirect PQC

The other major part of the PQC system is *indirect PQC*, defined as those costs not directly measurable in the company ledger but part of the product life-cycle PQC. They consist of three major categories:

- Customer-incurred PQC
- Customer-dissatisfaction PQC
- Loss-of-reputation PQC

Customer-incurred PQC appears when an output fails to meet the customer's expectations. Typical customer-incurred PQC are:

- Loss of productivity because a service was not performed on schedule
- Travel costs and time spent to hold a second meeting because the correct data were not available at the first meeting
- Overtime expended because input schedules were not adhered to
- Cost of rethinking (and possibly changing) a decision because the data used to make the decision were in error

Customer dissatisfaction is a binary thing. Customers are either satisfied or dissatisfied. Seldom will you find one who is in between.

The quality level of U.S. and European products hasn't suddenly dropped. In fact, it has improved. What has happened is that the customer expectation level has changed. Customers now require a much better product to satisfy their expectations and demands. The customer dissatisfaction level has moved, but in many companies the quality level has remained constant or has not kept pace with customer expectations. These companies may very well have been making parts to specifications, but the specifications were not good enough to keep their old customers, let alone attract new ones. Many of our business leaders understand that customer expectations are changing. For example, F. James McDonald, former president of General Motors, said, "At General Motors, we're proud of the gains we've made in quality and in the productivity associated with the improvements, and our customers are verifying the results of our efforts. But even as our own performance improves, the expectations of our customers continue to rise." Customer dissatisfaction PQC is measured in loss of market share.

Loss-of-reputation PQC is even more difficult to measure and predict than are customer dissatisfaction and customer-incurred PQC. Costs incurred because of loss of reputation differ from customer-dissatisfaction costs in that they reflect the customer's attitude toward a company rather than toward an individual product line. The loss of a good reputation transcends all product lines manufactured by a company.

While vice president of General Electric, Armand V. Feigenbaum developed the concept of quality cost in the early 1950s. In 1987 he wrote,

> Our original development of the concept and quantification of quality cost has had the objective of equipping men and women throughout a company with the necessary practical tools and detailed economic knowhow for identifying and managing their own costs of quality. These costs have successfully provided the common denominator in business terms both for managing quality as well as for communication among all who are involved in the quality process. Therefore, we have continued to develop, implement, and refine the cost of quality in companies the world over.

USING MEASUREMENT DATA

By themselves, all the data in the world, even when analyzed in the most sophisticated ways, accomplish nothing. In fact, data collection, data analysis, and data storage are activities that add no value until the data are used to control, inform, or improve a process. The worth of the whole measurement system rests in how the data are used. Never collect data that will not be used. It is wasted effort. One of the fundamental reasons for having a process measurement system is to identify and prioritize process improvement and change opportunities.

In the PIT's quest for perfection, the process evolves through two major states:

Level 1:   Meeting business expectations (targets)
Level 2:   Excelling at the job

Meeting Business Expectations (Level 1)

The PIT's first obligation is to bring all measurements up to the business targets. That objective can be further prioritized into:

Priority 1:   Meeting end-customer expectations (effectiveness)
Priority 2:   Meeting total process efficiency targets
Priority 3:   Meeting subprocess and activity business targets

The PIT should ensure that the process will consistently provide output that meets customer expectations. Once this is accomplished, the team should focus its efforts on the total process efficiency business target. When we talk about efficiency, we really are talking about cost. If you are not meeting your business efficiency targets, your company could be losing money by performing the activity, because the process is consuming too many resources compared to the value-added content of the process.

Once the total process is meeting its business targets, the PIT's attention should turn to the internal business targets at each subprocess, department, and activity. Often these improvement activities are conducted by small groups within the process, and the PIT only monitors progress.

Excelling at the Job (Level 2)

After the business target has been reached, the PIT is ready to embark on its quest for excellence (level 2). Level 2 activities are driven by the people performing the activity and are directed at the excellence of the activities being performed.

We are now ready to start down the long road to error-free performance. In most cases, the process will never have to go back to level 1 because the quality of output will stay ahead of customer expectations—but that does not mean that business targets do not need to be reassessed. Quite the contrary, it is very important that they are continuously reevaluated so that the PIT and the people working within the process (department improvement terms, or DITs) understand where the process is performing related to customer expectations.

In level 2, we strive to make the process better today than it was yesterday and better tomorrow than it is today. To accomplish this goal, we can use the quality ring (Figure 7.10). To start the evolution to error-free performance, the PIT should review the PIC indicators and select up to 10 of them to improve. Involve the employees performing the activities in setting challenge targets. Next, assign a subprocess improvement team (sub-PIT) or a DIT to each of the focus improvement areas. The challenge targets should be more stringent than the business tar-

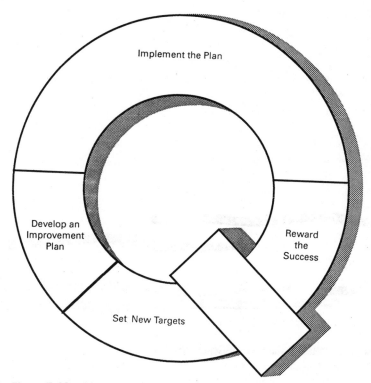

Figure 7.10   The quality ring.

gets. Develop a time-line action plan to improve the process performance. The result is a plan designed to improve the process by a specific amount in a specific number of months.

As the activity travels around the quality ring, the process should move into the implementation stage. It is important that the DIT or sub-PIT not merely stand by during this stage waiting for things to happen. The DIT or sub-PIT members should be the driving force. Develop frequent checkpoints to measure progress. Modify the plan as necessary when an activity is not as effective as originally estimated or as new data become available. Don't just wait until the end of the implementation stage and then suddenly realize that the process isn't going to meet the CT. Take action along the way to be sure that you will make it. If other areas are not meeting their commitments, insist that they provide a makeup schedule, and escalate the situation if you don't receive satisfactory answers. Remember, though, that the support groups are working on many processes and have several priorities. They should give first priority to meeting external customer expectations. Working to help the DIT or sub-PIT meet CTs is a lower priority.

Once you have met the CTs, recognize the DIT or sub-PIT for the accomplishment. The level of recognition should be commensurate with the accomplishment. It can range from a "job well done" at a meeting to a major financial award, depending on the magnitude of the contribution and the award system established by your organization.

Don't stop here. This is not the end; it is the start of a new beginning. After recognizing the team for meeting the CTs, schedule another meeting to develop new, more stringent targets. Develop a plan to meet them, implement it, recognize the team, and establish a new set of CTs. The teams should go around and around the quality ring until they reach the ultimate goal of error-free performance.

Ingredients Required to Correct Problems

Any long-range plan requires six critical ingredients to permanently eliminate problems. They are:

1. *Awareness.* The employees and management must be aware of the importance of eliminating errors and the cost impact that errors have on the business. In many companies, eliminating errors can reduce costs by over 30 percent.

2. *Desire.* Create a desire to eliminate errors. A lot more satisfaction is derived from doing the job right every time than from spending 25 percent of your time redoing it. No one wants to be wrong; give them a chance to do it right.

3. *Training in problem solving.* The individuals working to eliminate problems must be confident problem solvers. They should do more than just present problems; they should collect cost and supporting data. They must assemble alternative solutions to a problem and then select the best one.

4. *Failure analysis.* Develop a system to translate symptoms into a precise understanding of what caused the problem (failure mode). Without such data, many problems can be solved only by expensive and time-consuming trial-and-error methods. In many cases, interrelated failure modes take place at the same time. It often is impossible to complete the preventive action process without detailed failure analysis data because the failure disappears without identifying and correcting the true failure mode. Such problems come back to bite you again later on.

5. *Follow-up system.* Having a system to track problems and action commitments is essential. It should provide a means for evaluating the effectiveness of the corrective action.

6. *Liberal credit.* Liberally give credit and recognition to all who participate.

OPPORTUNITY CYCLE

Things go wrong. Everything is not perfect. Letters get lost, customers complain, we lose orders we should have won, and we miss design schedules. The list goes on. It is enough to discourage anyone, if all we do is look at the waste. What we should do is look at this as an opportunity to contribute to the success of our organization. It's like prospecting for gold. Each problem you solve is like finding a large gold nugget. Eliminating problems represents a major opportunity.

When you investigate each problem, go through the six distinct phases indicated in Figure 7.11. Each phase contains a number of individual activities. The total cycle consists of 25 different activities.

Phase 1—Opportunity Selection Phase
- Activity 1—Listing the problems
- Activity 2—Collecting data
- Activity 3—Verifying the problems
- Activity 4—Prioritizing the problems

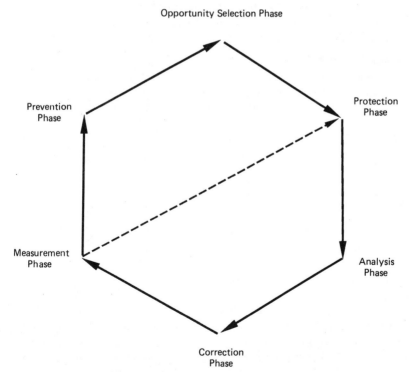

Figure 7.11 The opportunity cycle.

- Activity 5—Selecting the problems
- Activity 6—Defining the problems

Phase 2—Protection Phase

- Activity 7—Taking action to protect the customer
- Activity 8—Verifying the effectiveness of the action taken

Phase 3—Analysis Phase

- Activity 9—Collecting problem symptoms
- Activity 10—Validating the problem
- Activity 11—Separating cause and effect
- Activity 12—Defining the root cause

Phase 4—Correction Phase

- Activity 13—Developing alternative solutions
- Activity 14—Selecting the best possible solution
- Activity 15—Developing an implementation plan
- Activity 16—Conducting a pilot run
- Activity 17—Presenting the solution for approval

Phase 5—Measurement Phase

- Activity 18—Implementing the approved plan
- Activity 19—Measuring cost and impact
- Activity 20—Removing the protective action (installed in Phase 2)

Phase 6—Prevention Phase

- Activity 21—Applying action taken to similar activities
- Activity 22—Defining and correcting the basic process problem
- Activity 23—Changing the process documentation to prevent recurrence
- Activity 24—Providing proper training
- Activity 25—Return to Phase 1, Activity 1

SUMMARY

To measure is to understand, to understand is to gain knowledge, to have knowledge is to have power. The thing that sets humans apart from the other animals is our ability to observe, measure, analyze, and use this information to bring about change.

As our measurement systems improved, our standard of living improved. The whole industrial revolution was possible because our measurement systems improved to the point at which parts could be interchanged with each other. As measurement systems have improved, they have supported new product development. The present computer systems supporting our service industries and manufacturing support areas are highly dependent on very sophisticated measurement systems.

The manufacturing and product development measurement system has become very sophisticated, while the support areas and service industries have developed little. There is much to be gained from improving our business process measurement system. Through its use, we will gain increased knowledge. Often, we hear that companies cannot afford to measure their business processes. We contend that you cannot afford *not* to measure them.

ADDITIONAL READING

Feigenbaum, Armand V., *Total Quality Control*, 2d ed., McGraw-Hill, New York, 1983.

Harrington, H. James, *The Improvement Process*, McGraw-Hill, New York, 1987, chaps. 5 and 10.

————, *Poor-Quality Cost*, Marcel Dekker and ASQC Quality Press, Milwaukee, Wisconsin, 1988.

The Statistical Process Control Manual, Harrington, Hurd & Rieker, San Jose, CA, 1989.

Talley, Dorsey J., *Total Quality Management—Cost and Performance Measures*, ASQC Quality Press, Milwaukee, Wisconsin, 1991.

CHAPTER

8

Process Qualification

INTRODUCTION

Many people never start down the long, long road to continuous improvement because they see no end in sight. They reason that management, in a continuous improvement environment, is never satisfied. It will always demand more and better. Many people consider this a win for management, and a loss for themselves.

How can we as managers overcome such negative employee opinions? The key is to establish win points (intermediate goals) along the road to perfection. Consider a goal of walking across the United States. It is such a huge undertaking that few will start and even fewer will ever finish. Those who do will focus their attention, not on walking across the entire country, but on walking from San Francisco to Sacramento (a win point). Then, after a good night's sleep, a hearty meal, and a long, hot bath in a comfortable hotel, they set their sights on walking from Sacramento to Reno, with another reward when they arrive there.

Becoming the best is a lofty, difficult objective. To have the best business processes should be everyone's goal, but we need milestones along the way to show us that we are making progress. That's what business process qualification is all about. It provides milestones and recognition points for the process improvement teams (PITs).

MANUFACTURING PROCESS QUALIFICATION

In the 1960s, progressive companies introduced formal process qualification procedures for their manufacturing processes. For example, an IBM technical report entitled *Process Qualification—Manufacturing's Insurance Policy* discussed how to evaluate and control manufacturing processes, from the development laboratory's bench process, to the pilot line, and into the highly sophisticated, automated production process. It outlined a four-level qualification process used to evaluate manufacturing processes prior to product shipment to external customers. At each progressive stage, new expectations are set for:

- Documentation
- Test and/or process equipment
- Software
- Manufacturing operations
- Product performance

During the process qualification procedure, the process windows, equipment capability, process control points, training specifications, throughput limitations, and manufacturing cycle time are defined and verified.

Manufacturing process qualification normally is conducted before the first product is shipped to a customer. It systematically helps the process concept to evolve, to maximize its efficiency and effectiveness, and to ensure product performance.

The same thought pattern can aid any organization in improving its business processes. Concepts like capability, repeatability, and reliability apply equally well to this situation. Let's define the key terms.

Certification. Certification is applied to a single activity or piece of equipment. When an evaluating group is confident that an individual activity, operation, person, or piece of equipment will provide output that meets the next activity's requirements using the documentation, the item can be certified.

Qualification. Qualification involves evaluating a complete process, consisting of many individually certified activities, to determine whether the process can perform at the appropriate level when the activities are linked together. In addition, the process must demonstrate that it can repeatedly produce products and/or services on time, at the appropriate cost, that meet customer expectations on an ongoing basis.

COMPARISON OF MANUFACTURING AND BUSINESS PROCESS QUALIFICATION

Manufacturing and business process qualification strategies have important similarities, but also some notable differences.

As in manufacturing process qualification, our first need in business process improvement (BPI) is to establish qualification levels and procedures for the total process. Not only will we qualify the total process, we will certify activities or subparts of the process. Certification activities are a major part of the business process qualification strategy. For example, we could certify that the software is adequate for a given job, or we could certify that an office employee has been trained to perform his or her job at the required level. Unfortunately, at the present time, most organizations do not certify individual activities in their business processes. However, as business processes become increasingly important to improving an organization's effectiveness, certification will be used more frequently. In fact, certification will become as common in business process qualification as it is in manufacturing.

Manufacturing and business process qualifications also share similar goals and activities. For example, we will require from a business process proof that it is *capable* of generating the output we expect from it in terms of timing, content, service, etc. We will also require that the business process qualification measure repeatability and effectiveness of output. Repeatability is a key requirement of manufacturing. If you buy a recorder, a television, or a car, you expect it to be as reliable as the ones made before and after it. You expect it to start every time. You do not accept variation. Why should this be any different for a business process? Why can't billing, purchasing, sales, and other standard customer service activities be just as predictable and repeatable as manufacturing processes? McDonald's success is largely due to its ability to consistently provide the same quality of service and product every day, anywhere in the world. Yes, service customers want—no, they demand—repeatability of output.

In addition, both manufacturing and business process qualification must be able to achieve effective mass production. It's not enough when a process operates perfectly only on a computer or on the drawing board. It must be fully operational and effective when it is required to process hundreds, thousands, or millions of transactions—both when things are slow and when the pace is hectic.

The final similarity relates to documentation. Business processes, as we have stated earlier, usually are not well documented. However, as in manufacturing, business process documentation will be used for process follow-up and personnel training. If each time a business process is used, it is executed in a different way, improvement is difficult, if not impossible. Proper documentation is the key to standard practices and quality. First you need to standardize, then to start to improve.

Are there differences between manufacturing and business process qualification strategies? At first glance, one might say there is no meaningful difference between the two. Indeed, if we were to design a new business process from scratch, it would be easy to follow all the qualification steps required in the manufacturing qualification strategy. Unfortunately, most of today's business processes were designed years, if not decades, ago. Consequently, we cannot start over but need to build on the existing processes. This means that we cannot qualify the original design of the process, as in a manufacturing application. We must requalify an existing design and adapt it to new customer or business requirements while it is functioning.

IS BUSINESS PROCESS QUALIFICATION NECESSARY?

Manufacturing process qualification guarantees that the process design provides customers with acceptable products. As manufacturers, we wish to demonstrate consistency and quality. Each new process puts our business reputation at risk. One bad process can destroy years of hard work. Customers remember the bitterness of poor performance long after the sweetness of outstanding service has faded.

Because business processes largely service internal customers, it is easy to lose sight of their importance to our overall business survival. This is where process qualification helps, by motivating us to take the first steps toward continuous improvement. Whether office professionals, clerical employees, or middle managers, people love to be recognized for their efforts and are stimulated by public acknowledgment. Process qualification provides a measurement system that instills a sense of pride within each team.

Process qualification supports the goals set forth in the beginning of the BPI effort. There is no better way of showing how serious we are than by installing a systematic business process qualification strategy.

Process qualification will motivate the PIT and keep that motivation high. You must continuously improve. If you are not improving, if you are standing still, you are not holding your own. In fact, you are sliding backward, because your competition is improving. You can't earn this week's paycheck with last week's press clippings. Qualifying a process typically includes these steps:

- The PIT evaluates the process using the appropriate requirements list.
- The PIT leader requests qualification level change.
- The BPI champion (czar) reviews the process status.
- The PIT prepares a process status report and sends it to the review committee. The EIT often serves as the review committee.
- The PIT presents the process change data to the review committee.
- The review committee chairman issues the process qualification change letter.
- The review committee rewards the PIT for its accomplishments.

BPI LEVELS

A six-level qualification process can provide an effective structure and guide for BPI activities. These levels lead the PIT from an unknown process status to the ultimate best-of-breed classification.

Level	Status	Description
6	Unknown	Process status has not been determined.
5	Understood	Process design is understood and operates according to prescribed documentation.
4	Effective	Process is systematically measured, streamlining has started, and end-customer expectations are met.
3	Efficient	Process is streamlined and is more efficient.
2	Error-free	Process is highly effective (error-free) and efficient.
1	World-class	Process is world-class and continues to improve.

Until the BPI methodology has been applied, all business processes are considered to be at level 6. As the process improves, it progresses logically up to level 1. This enables the organization to quickly evaluate the process status (see Figure 8.1).

A quick look at a business process overview chart provides the status of an organization's business processes. In the case of accounts payable, for example, the organization has decided that it does not need to benchmark the business process. In this case, the executive improvement team (EIT) is satisfied to have the process under control and performing to standards. In the case of order entry, the organization wanted to benchmark but set a target to be as good as, but not better than, the best. These types of decisions are usually based on business factors and priorities. As business conditions and priorities change, the decision not to be the best in all critical business processes should be readdressed.

Business process	Levels					
	6	5	4	3	2	1
Order entry	X	X				NR
Accounts payable	X				NR	NR
Customer complaint handling	X	X	X	X		
Special product price request		X				
Engineering change release	X	X				
Customer-requested information	X	X				
NR = not required						

Figure 8.1   Business process overview chart.

All processes in all organizations may not need to progress through all six levels. Often, there are considerable costs involved in becoming the best. In most cases, organizations have many business processes that need to be improved. Because of the magnitude of this job, it may be wise to bring some of the processes under control and then direct the limited resources to another critical business process. Once all the critical processes are under control, PITs can be assigned to bring the most critical processes up to level 1 (world-class). Often, the people using the process will make major improvements in it as a result of their continuous improvement activities, while the PIT is working on another process. It is important that the process owner keep the process responsibility so that suboptimization does not occur while the PIT is working on another process. When the EIT decides that something less than a world-class performance level is acceptable, it should communicate this information to the PIT immediately.

Ideally, this decision should be made prior to forming the PIT. When management decreases its expectations late in the process cycle, it can negatively impact the PIT's morale by being interpreted as management's loss of faith.

DIFFERENCES AMONG BUSINESS PROCESS LEVELS

To determine whether the process has evolved to the next level, eight major change areas should be addressed:

- End-customer-related measurements
- Process measurements and/or performance
- Supplier partnerships
- Documentation

- Training
- Benchmarking
- Process adaptability
- Continuous improvement

Each of the following five sections of this chapter will define the requirements to move from one qualification level to another.

The following definitions will help you understand the changing expectations that must be met to change qualification levels:

Requirements.   What the customer must be supplied with

Expectations.   What the customer would like to have to do his or her job the best; what the customer thinks can reasonably be provided or that can be obtained from a competitor

Desires.   The customer's wish list; what it would be nice to have but is not essential

REQUIREMENTS TO BE QUALIFIED AT LEVEL 5

Qualification level 5 signifies that the process design is understood by the PIT and is operating to the prescribed documentation.

All processes are classified as level 6 until sufficient data have been collected to determine their true status. Normally, processes move from qualification level 6 to qualification level 5. To be qualified at any level, all the criteria in each of the eight major change areas (for example, supplier partnerships, process measurements and/or performance) must be met or exceeded. Those for level 5 are:

End-customer-related measurements
- Measurements reflect the end customer's view of the process.
- End-customer requirements are documented.
- End-customer feedback system is established.
- End-customer effectiveness charts are posted and updated.

Process measurements and/or performance
- Overall effectiveness and efficiency are measured and posted where they can be seen by employees.
- Effectiveness and efficiency targets are set.
- Process operational and/or control weaknesses are evaluated and meet minimum requirements.

Supplier partnerships
- All suppliers are identified.

Documentation

- Process is defined and flowcharted.
- Flowchart accuracy is verified.
- Documentation is followed.
- PIT members and process owners are named.
- PIT mission is documented.
- Process boundaries are defined.

Training

- PIT is trained in the basic tools and the fundamental BPI tools.
- In-process training needs are evaluated and documented.
- Resources are assigned to support training needs.

Benchmarking

- Not required.

Process adaptability

- Not required.

Continuous improvement

- Basics of BPI are in place.
- All major exposures are identified, and action plans are in place.
- A detailed plan to improve the process to level 4 is agreed to and funded.

REQUIREMENTS TO BE QUALIFIED AT LEVEL 4

When a process evolves to qualification level 4, it is called an *effective process*. Processes qualified at level 4 have a systematic measurement system in place that ensures end-customer expectations are met. The process has started to be streamlined.

To be qualified at level 4, the process must be able to meet all the requirements for qualification level 5, plus the following requirements:

End-customer-related measurements

- End-customer requirements are met.
- End-customer expectations are documented.

Process measurements and/or performance

- Overall effectiveness targets are met, and challenge targets are established by the PIT.
- Poor-quality cost (PQC) measurements are developed.
- Some internal efficiency measurements are established.
- Internal effectiveness measurements and targets are 50 percent complete and posted.

- Overall process cycle time and cost are defined.
- No significant effectiveness, efficiency, or control exposures exist.
- Substantial improvement activities are under way.

Supplier partnerships

- Meetings are held with critical suppliers, and agreed-to input requirements are documented.
- Feedback systems to critical suppliers are in place.

Documentation

- Process is flowcharted, and documents are updated.
- Overall process is fully documented.
- Documentation of subprocesses starts.
- Readability is evaluated.

Training

- In-process job training procedures are developed for all critical activities.
- People are assigned to conduct job and process training.
- PIT is trained in statistical process control.

Benchmarking

- Plan exists to benchmark end-customer requirements.

Process adaptability

- Data are collected that identify problems with present process adaptability.

Continuous improvement

- Process is operational, and control weaknesses are assessed and deemed containable.
- A plan for improving the process to level 3 is prepared, approved, and funded.
- The process philosophy accepts that people make mistakes, provided everyone works relentlessly to find and remove causes of errors.

REQUIREMENTS TO BE QUALIFIED AT LEVEL 3

When a process evolves to qualification level 3, it is called an *efficient process*. Processes qualified at level 3 have completed the streamlining activities, and there has been a significant improvement in the efficiency of the process.

To be qualified at level 3, the process must be able to meet all the requirements for qualification levels 5 and 4, plus the following:

End-customer-related measurements

- End-customer expectations are met.

- Challenge targets are set by the PIT.

Process measurements and/or performance

- There is a significant improvement in PQC.
- Internal effectiveness and efficiency measurements are in place and are posted, with targets set by the affected areas.
- There is a significant reduction in cycle time and bureaucracy.
- Overall efficiency targets are met.
- Most measurements show an improvement trend.
- Key process control points are identified.
- Tangible, measurable results are realized.

Supplier partnerships

- Meetings are held with all suppliers, and agreed-to input requirements are documented.
- All critical suppliers meet input requirements.

Documentation

- Subprocesses are documented.
- Training requirements are documented.
- Software controls are in place.
- The readability level of all documents is at a grade level less than the minimum education of the people using them.
- Employees understand their job descriptions.

Training

- All people performing critical jobs are trained in the new procedures, including job-related training.
- In-process job training procedures are developed for all activities.
- Plans are in place to train all employees who are part of the process in team methods and problem-solving tools.
- PIT understands one or more of the BPI 10 sophisticated tools.
- All employees in the process receive training on the total process operation.

Benchmarking

- End-customer requirements are benchmarked.
- Plan exists to benchmark critical activities.
- Plan exists to benchmark the process.

Process adaptability

- Employees are trained to distinguish how far they can deviate from the established procedures to meet a customer's special needs.
- Future process change requirements are projected.
- A proactive internal and external customer complaint system is established.
- The end customer reviews the process change plan and agrees that it meets his or her needs over the strategic period.

Continuous improvement
- A plan to improve the process to level 2 is developed, approved, and funded.

REQUIREMENTS TO BE QUALIFIED AT LEVEL 2

When a process has evolved to qualification level 2, it is called an *error-free process*. Processes qualified at level 2 are highly effective and efficient. Both external and internal customer expectations are measured and met. Rarely is there a problem within the process. Schedules are always met, and stress levels are low.

To be qualified at level 2, the process must be able to meet all the requirements for the previous qualification levels, plus the following requirements:

End-customer-related measurements
- End-customer expectations are updated.
- Performance for the last 6 months never fell below end-customer expectations.
- The trend lines show continuous improvement.
- World-class targets are established.
- End customers are invited to regular performance reviews.
- End-customer desires are understood.

Process measurements and/or performance
- All measurements show an improvement.
- Benchmark targets are defined for external customers and critical in-process activities.
- In-process control charts are implemented as appropriate, and the process is under statistical control.
- Feedback systems are in place close to the point at which the work is being done.
- Most measurements are made by the person doing the job.
- There is tangible and measurable improvement in the in-process measurements.
- No operational inefficiencies are anticipated.
- An independent audit plan is in place and working.
- The process is error-free.

Supplier partnerships
- All supplier inputs met requirements for the last 3 months.
- Regular meetings are held to ensure that suppliers understand the changing needs and expectations of the process.

Documentation
- Change level controls are in place.
- Documents are systematically updated.

Training
- All employees in the process are trained and scheduled for refresher courses.
- Employee evaluation of their training process is complete, and the training meets all employee requirements.
- Team and problem-solving courses are complete. Employees are meeting regularly to solve problems.

Benchmarking
- Process is benchmarked, and targets are assigned.
- PIT understands the keys to the benchmark organizations' performance.

Process adaptability
- Employees are empowered to provide the required emergency help to their customers and are measured accordingly.
- Resources are committed to satisfy future customer needs.
- Process adaptability complaints are significantly reduced.

Continuous improvement
- The process philosophy evolves to the point at which errors are unacceptable. Everyone works relentlessly to prevent errors from occurring even once.
- Surveys of the employees show that the process is easier to use.
- Plans to improve the process to level 1 are prepared, approved, and funded.

REQUIREMENTS TO BE QUALIFIED AT LEVEL 1

Qualification level 1 is the highest qualification level. It indicates that the process is one of the 10 best processes of its kind in the world, or it is in the top 10 percent of like processes, whichever has the smallest population.

Processes that reach qualification level 1 are called *world-class processes*. Processes qualified at level 1 have proved that they are among the best in the world. These processes are often benchmark target processes for other organizations. As a rule, few processes in an organization ever get this good. Processes that reach level 1 truly are world-class and continue to improve so that they keep their world-class process status.

To be qualified at level 1, the process must be able to meet all the requirements for the previous qualification levels, plus the following:

End-customer-related measurements
- End-customer expectation targets are regularly updated and always exceeded.
- World-class measurements are met for a minimum of 3 consecutive months.
- Many of the end-customer desires are met.

Process measurements and/or performance
- All measurements exceed those of the benchmark company for 3 months.
- Effectiveness measurements indicate that the process is error-free for all end-customer and in-process control points.

Supplier partnerships
- All suppliers meet process expectations.
- All suppliers meet process requirements for a minimum of 6 months.

Documentation
- All documents meet world-class standards for the process being improved.

Training
- Employees are regularly surveyed to define additional training needs, and new training programs are implemented based on these surveys.

Benchmarking
- Ongoing benchmarking plan is implemented.

Process adaptability
- In the last 6 months, no customers complained that the process did not meet their needs.
- Present process handles the exceptions better than the benchmark company's process.

Continuous improvement
- An independent audit verifies world-class status.
- Plans are approved and in place to become even better.

It is important to note that the goal for all level 1 processes is to go beyond world-class to become the *best-of-breed process*. Although some processes become best-of-breed for short periods of time, it is very difficult to stay no. 1. It requires a great deal of work and creativity, but the personal satisfaction is well worth it.

SCHEDULING A BUSINESS PROCESS REVIEW

In Chapter 2, the EIT was designated as the leading committee for all BPI activities. Although this remains true, it may be inappropriate to have all the PIT teams report directly to the EIT.

As a rule of thumb, the PIT should report to the management level that encompasses the process under study. For example, if the process boundaries are such that the process covers more than one vice president or director, this PIT should report to the EIT chaired by the CEO of the organization. On the other hand, if the process is contained within marketing, the PIT could report to the vice president of marketing. In turn, the vice president should report PIT status to the EIT.

When the PIT's data indicate that the process has progressed to the next qualification level, the process owner should ask the BPI champion (czar) to schedule a process qualification review meeting.

ASSIGNING QUALIFICATION LEVELS

When the BPI champion receives a request to change a process qualification level, he or she should meet with the process owner to review the process data to ensure they are complete. If the process is ready for a formal review, the BPI champion should schedule a process qualification review meeting with the EIT or with the appropriate management team. Prior to the meeting, the PIT should submit a formal process status report containing:

- Process backup data (i.e., process name, PIT mission, PIT members, process scope)
- Status of all measurements
- Process flowchart
- Documented current status compared to the requirements to change to the next level
- Improvements made since the latest level change
- Problems or exposures solved
- Unsolved problems or exposures
- Plan to improve the process to the next qualification level

At the process qualification review meeting, the PIT should compare the current process to the requirements for the new qualification level. Frequently, the PIT will invite satisfied customers to attend these meetings to testify as to the process status. Once the review committee verifies that all requirements have been met, the committee should issue a formal qualification letter stating that the process has evolved to the appropriate level.

RECOGNITION AND REWARD PROCESS

It is strongly suggested that the EIT establish a recognition and reward process to acknowledge the PIT each time its business process is qualified at a higher level. The nature of the reward should increase as the qualification level becomes more difficult to obtain. A typical reward structure is:

Level		Reward
From	To	
6	5	Article in the organization's newsletter.
5	4	Lunch for the PIT with the EIT. Present a special recognition gift.
4	3	Dinner for two for each PIT member and a guest. Also, a check equal to the individual's share of 10 percent of the documented savings since the PIT was formed.
3	2	Formal dinner with the EIT. Give a special gift to each PIT member.
2	1	Yearly recognition event at a vacation resort for the PIT team and their guests. The first year that the business process is recognized as world-class, give all PIT members a special contribution award of $10,000 cash, or 50 percent of 1 year's documented savings, whichever is less.

SUMMARY

There is only one best of anything. There are some near bests, and many that are average, but 50 percent of the processes are always below average. Don't expect all processes in your organization to reach level 1. In fact, many may never even reach level 2—but that should not prevent you from trying to be better than you are. The only acceptable standard is perfection—in ourselves, in our work, and in our business processes. We should not stop short of perfection. As Thomas J. Watson, Sr., first president of IBM, put it, "It is better to aim at perfection and miss, than to aim at imperfection and hit it."

Even when your process becomes the best-of-breed, you must continue to improve. You can't relax for a minute because when you stop improving, you start slipping backward. It takes an active continuous improvement process just to maintain the level, because people change, systems change, and customer needs change, so the process has to change and improve all the time. If some day you improve your process to the level of perfection, take something out of it to upset the status quo so you can start the next generation of improvement.

Benchmarking Process

INTRODUCTION

At this point, you have spent a lot of time understanding and characterizing your process. You have developed ways to measure its effectiveness and efficiency. And soon you will be making significant improvements in both of these key business areas. You can do a lot to bring about change. You can eliminate bureaucracy, simplify the process, and reduce the time it takes to use it. But how much change is really necessary or possible? In short, what is the world-class standard for your process?

This is the point at which people usually ask, "Is it possible for our process to be better? If so, what can we do to make it better? Don't give us that old stuff about continuous improvement; we want specific implementation advice, not just more theory. Where do we go from here? We have used up all of our ideas; maybe we have gone as far as we can go with this process."

Chances are that there still is "gold to mine in them thar hills." To find it, what you need to do now is to look outside your own location at other, similar processes within your own organization and at outside organizations as well. Your purpose is to understand what they are doing and to use this combined experience and knowledge to help develop your process even further. This act of systematically defining the best systems, processes, procedures, and practices is called *benchmarking*.

Becoming the very best in any field is a difficult and lonely road to travel. Once you reach your goal, there is only one direction to go: down. When you are the best, you are envied, undermined, frequently criticized, and expected to outperform the competition no matter what the circumstances. Why, then, do so many people, teams, and organizations want to be recognized as the very best? The reasons are simple. Excellence brings:

- Satisfaction
- Recognition
- Higher rewards
- Customers
- Respect
- Power
- Money

To become the very best you must:

- Know yourself, your strengths, and your limitations
- Recognize and understand the leading organizations in the area in which you hope to excel
- Use the best processes available
- Build on these processes to create even better ones

The benchmarking process (BMP) helps you to know yourself, understand your competition, define the best processes, and integrate them into your organization. This chapter reviews the BMP and discusses how to apply it to the business process improvement (BPI) concept to develop world-class goals for your process. It also defines the activities that need to be changed to upgrade your process to meet these world-class standards.

THE BMP OVERVIEW

David T. Kearns, chief executive officer of Xerox Corporation, defines benchmarking as "the continuous process of measuring products, service, and practices against the toughest competition or those companies recognized as industrial leaders." Benchmarking is a never-ending discovery and learning experience that identifies and evaluates best processes and performance in order to integrate them into an organization's present process to increase its effectiveness, efficiency, and adaptability. It provides a systematic way to identify superior products, services, processes, and practices that can be adopted or adapted to your environment to reduce costs, decrease cycle time, cut inventory, and provide greater satisfaction to your internal and external customers.

Benchmarking often is viewed as simply purchasing competitive products to compare them with the ones manufactured by the organization. This is only one type of benchmarking and not the one that will be discussed in this chapter. Competitive product disassembly is only a small part of the larger BMP. We will be discussing how to define and understand the best business practices and processes to enable your organization to provide superior products and service at reduced cost.

The BMP is a lot like a detective story, and the person doing the benchmarking operates a lot like a detective. He or she must search through the many clues available in the public domain to find leads and then follow up on these leads to identify and understand the truly world-class processes. It can be an exciting and enlightening adventure.

WHY USE BENCHMARKING?

The two primary reasons for using the BMP are goal setting and process development.

Every person, process, and organization need goals to strive for. Without them, life is unrewarding, and we drift on a sea of confusion. We all want to improve. No one likes to be average. In the past, goals usually were based on the organization's or the process's past performance. There was little correlation between our goals and the ultimate standard of excellence. Occasionally, our goals exceeded what is achievable, but more commonly they fell far below what had been, or could be, achieved. As a result of setting low goals for ourselves, we enjoyed a false sense of accomplishment. We stopped trying to improve because we so easily met the low standards we set for ourselves. This prevented many individuals, processes, and organizations from maturing to their full potential. Because it provides a means for setting challenging targets and attainable goals, benchmarking is the antidote to this self-imposed mediocrity.

Even more important, the BMP provides a way to discover and understand methods that can be applied to your process to effect major improvements. That is the unique value of the BMP. It not only tells you how good you need to and can be, it also tells you how to change your process to get there.

The BMP comprises two parts: measurement (the what) and product and/or process knowledge (the how). You need to address both functions. What good is knowing how good world-class is if you do not know how to improve your process to obtain it? Knowing that you are bad, but not being able to improve, just discourages everyone. Similarly, what good is a new process idea if you do not know whether it will have a positive impact on your process? You need to design your BMP to provide both the what and the how.

The What

Measurements are absolutely crucial. If you cannot measure, you cannot control. If you cannot control, you cannot manage. It's as simple as that. Quantitative data are absolutely essential ingredients in becoming, and staying, world-class.

As critical as measurements are, it would seem that everyone would know just what needs to be measured. Unfortunately, this is not the case. In fact, in most instances it is just the opposite, and this is particularly true when you talk about business processes. The BMP should measure things such as:

- How much
- How fast
- How good
- When
- Where
- How long
- Size, shape, form, and fit

Although most product-type measurements are physical in nature, most process measurements are effort, cost, and time related. Consequently, it often is best to establish ratio measurements (e.g., return on investments, returns per year, unit costs, productivity rates) rather than actual values. The use of ratios allows data exchange without disclosing absolute values or production rates. This encourages free exchange of information between organizations.

The How

Another real advantage of the BMP is that it provides you with insights into how others have become the best. This aspect focuses on discovering how world-class organizations developed their processes and systems to ensure superior performance. At this juncture, we seek and analyze the how-to's, the knowledge, the ways, the processes, and the methods responsible for making an organization, a process, or an activity the best of its kind. We then apply this knowledge to our process, adapting it to meet the unique requirements of our products, employees, customers, and organization's personality.

WHAT WILL BENCHMARKING DO FOR YOU?

Benchmarking requires a lot of work, and once you start, it is an ongoing process to keep it updated. Why, then, should any organization even begin? Because the benefits of benchmarking far outweigh the effort and expense. Benchmarking:

- Provides you with a way to improve customer satisfaction
- Defines best applicable processes
- Improves your process
- Helps eliminate the "not-invented-here" syndrome
- Identifies your competitive position
- Increases the effectiveness, efficiency, and adaptability of your processes
- Transforms complacency into an urgent desire to improve
- Helps set attainable, but aggressive, targets
- Increases the desire to change
- Allows you to project future trends in your industry
- Prioritizes improvement activities
- Provides your organization with a competitive advantage
- Creates a continuous improvement culture
- Improves relationships and understanding between benchmarking partners

HISTORY OF THE BMP

During the 1960s, IBM realized that its costs could be reduced significantly, and the quality of its process-sensitive products improved, if its worldwide locations adopted the best existing practices. As a result, a corporate procedure was written requiring all process-sensitive products to be manufactured using compatible processes. This launched a corporatewide effort to have common practices at all locations or, if that proved impractical, to have compatibility between common processes. The determination to identify the best manufacturing processes gave IBM a significant international competitive advantage.

In the late 1970s, in a similar move, Xerox decided to compare its U.S. products with those of its Japanese affiliate, Fuji-Xerox. Xerox was shocked to discover that Fuji was *selling* copiers at a price equivalent to what it cost U.S. Xerox simply to *manufacture* the copiers. This discovery spearheaded a successful program to reduce costs in the U.S. manufacturing process. Based on the success of this pilot program, Xerox management later incorporated benchmarking as a key element in its corporatewide improvement efforts. The formal initiation of the program began around 1983. Through this program, benchmarking took on new dimensions, and benchmarking techniques were applied to support processes as well as product processes.

Today, many organizations use benchmarking to help drive their continuous improvement efforts. For example, Motorola cites benchmarking as one of the major tools powering its improvement process, for which it was awarded the Malcolm Baldrige Award in 1988.

TYPES OF BENCHMARKING

Essentially, there are four generic types of benchmarking:

1. Internal
2. Competitive
3. World-class operations
4. Activity type

Internal Benchmarking

Internal benchmarking involves looking within your own organization to determine whether other locations are performing similar activities and to define the best practices observed. This type of benchmarking is the easiest to conduct because there are no security and/or confidentiality problems to overcome. In almost all cases, this type of benchmarking should be undertaken first, since it is inexpensive to conduct and provides detailed data. Even better, you frequently can borrow experienced personnel from other locations to help improve your own process.

Competitive Benchmarking

Competitive benchmarking, which is also known as *reverse engineering*, requires investigating a competitor's products, services, and processes. The most common way to do this is to purchase competitive products and services and then analyze them to identify competitive advantages.

Originally, some organizations were apprehensive about the ethics involved in this process. Today, however, few organizations can expect to compete without a thorough understanding of the competition's products and services. Even organizations like IBM, Xerox, General Motors, and Hewlett-Packard make effective use of this method. In fact, Ford Motor Company carefully disassembles competitive cars and places them piece by piece in rows to compare their designs and assembly methods.

In addition, competitive products are extensively tested to identify their strengths and weaknesses and to generate performance profiles. A thorough, competitive BMP also reviews key information not directly related to the product. Close examination of the packaging, operating manuals, service instructions, and delivery methods can provide a great deal of valuable information.

World-Class Operations Benchmarking

World-class operations benchmarking extends the BMP outside of the specific organization, and its direct competition, to involve dissimilar industries. Many business processes are generic in nature and application (e.g., warehousing, supplier relations, service parts logistics, advertising, customer relations, and hiring) and can provide meaningful insights despite being used in an unrelated industry. Benchmarking dissimilar industries enables you to discover innovative processes not currently used in your particular product types that will allow your process to become the best-of-breed.

Setting some limitations on the organizations to be benchmarked will provide meaningful results at less cost, however. Some restrictions you may want to consider are:

- *Customer requirements.* High-quality and high-reliability or low-quality and one-time usage
- *Product characteristics.* Size, shape, weight, environment, etc.
- *Output usage.* (In this case, consider the broad industrial categories, not the specific products.) Grocery industry, office products industry, electronics industry, transportation industry, etc.

The BMP can be applied to a product, a process, a subprocess, or even an individual activity. Xerox has actively applied the BMP at all levels. Figure 9.1 lists some of the benchmark organizations Xerox has identified.

Organization	Product or process benchmark category
Canon	Copiers
DEC	Work stations
L. L. Bean	Warehouse operations
General Electric	Information systems
Deere	Service parts logistics
Ford	Assembly automation generic processes
Federal Reserve	Bill scanning
Citicorp	Document processing

Figure 9.1 Typical benchmark organizations identified by Xerox. (Robert C. Camp, Benchmarking—The Search for Industry Best Practices That Lead to Superior Performance, ASQC Quality Press, Milwaukee, WI, 1989, p. 62)

Activity-Type Benchmarking

Activity-type benchmarking is directed at process steps or specific process activities, such as engineering change control, order entry, invoicing, accounts payable collection, payroll, or recruiting, and it transcends industries.

THE BMP

Very simply, the BMP involves:

- Deciding what will be benchmarked
- Defining the processes to compare
- Developing measurements to compare
- Defining internal areas, and external companies, to benchmark
- Collecting and analyzing data
- Determining the gap between your process and the best process
- Developing action plans, targets, and measurement processes
- Updating the benchmarking effort

The six distinct phases of the BMP are:

1. Process design (planning)
2. Internal data collection
3. External data collection
4. Data analysis
5. Process upgrading
6. Periodic reassessment

What Should Be Benchmarked?

As a starting point, benchmark all your customer interfaces. Don't limit your benchmarking to products. Your organization's customer reputation is based on every contact your customer has with anything relating to your organization. This includes the contract trucker who delivers your product to a customer's warehouse, the way you answer your telephones, the accuracy of your bills, the responsiveness of your sales force, or the effectiveness of your service team. In addition, you should benchmark all your critical business processes, as well as key subprocesses and activities.

Process Design and Planning

A well-designed plan will save you a lot of time and headaches and will provide you with much better results. When an organization starts talking about benchmarking, the impulse is to run out and visit organizations that could have better processes. While this is something that should be considered, it is not the starting point.

The first step is to thoroughly understand yourself. If you have been involved in BPI activities, you are in a much better position to start the BMP. You have flow-diagramed your processes, you have established effectiveness and efficiency measurements, and you understand the interrelationships and the dependencies of your processes. All this will help you develop your BMP.

Key Processes and Systems Comparison

The next phase of the BMP is to closely evaluate the elements comprising the process being benchmarked. Identify those that:

- Have weaknesses within them
- Have a high potential for improvement
- Are sources of delay
- Take a large portion of the total effort
- Are the source of problems

Highlight these focus activities in the benchmarking study.

Measurements

Keep the list of key measurements you wish to benchmark as simple as possible. Do not use special terms uncommon to your industry. Analyze the type of data sources commonly available in your industry, and base your measurement requirements on this information. You have the most control of your own information, and it is often easier to reformat it than to get your benchmarking partners to provide data in your format. At this point, do not worry about the confidentiality of the measurements. At a later date, however, this subject will become very important.

Remember that ratio-type measurements always are best for the BMP. Effectiveness and efficiency measurements are key.

Benchmarking Target Locations and/or Organizations

The BMP generally starts by analyzing the best of the organization's internal operations. Many organizations have parallel operations performed at the same locations or at different locations throughout the world. We recommend beginning the process by assembling and sharing information about these parallel activities. These parallel operations may decide to form a team to develop, implement, and share the knowledge and cost of the BMP. Often, a great deal of information can be obtained by studying internal operations that, while not exactly the same, use similar methods (e.g., customer service problem reporting, in-process quality assurance reporting or order processing, and purchase requests processing). Additionally, internal experts in the benchmark areas, who are close to the technology and probably are involved in outside society activities, usually can provide the names of potential benchmarking partners and key contacts within these organizations. This list can be refined during the data collection and analysis stage.

Data Collection

The two ways to obtain data include retrieving published data (completed research) that are in the public domain (e.g., books, magazine articles, technical reports) and conducting original research (e.g., interviews, site visits, surveys). You probably will need to use both methods in developing your database. Evaluate each data source carefully for the following:

- Reliability and accuracy
- Availability
- Cost
- Coverage
- Timeliness
- Usefulness
- Usability
- Source
- Level of back-up

Your data collection system first should focus on collecting internal data to thoroughly understand how the processes within your organization are working before you approach external benchmark organizations. Information may be acquired in many ways, through:

- Internal experts
- Searches of the literature
- Professional and trade associations
- Consultants
- Universities
- External experts
- Surveys
- Location visits
- Focus groups

This is the order in which you normally would search out information. Do not rule out the use of a third party to do the research work for you, however. A third party has the advantage of being able to keep individual responses anonymous.

Data Analysis

This is a critical phase of the BMP because you must organize masses of numbers and statements into coherent, usable information to direct all your future activities. The success or failure of the BMP depends on how well the reams of collected data are translated into actionable information.

The measurement data provided you with indicators of the best practices, procedures, and processes. As you compare the benchmark data with your process, you may find that you are the best (world-class), the same, or worse. If you are the best, congratulations. If your comparison is negative or the same, an opportunity exists to improve by studying another organization's or location's process.

Two types of data collected and used in the BMP include qualitative data (word descriptions) and quantitative data (numbers, ratios, etc.). There has been much debate over which to collect first and how to use each type of data. In reality, your data collection strategy should be designed to collect both types of data as opportunities present themselves. A quantitative data matrix should be developed and filled out during the data collection cycle. This data matrix should highlight the parts of the process requiring additional data and study. It is best to have this matrix completed as thoroughly as possible before doing surveys or visiting organizations (see Figure 9.2).

	Organization			
	A	B	C	Ours
1. Average days to bring new employees on board	45	65	20	45
2. Number of approvals required	4	6	3	4
3. Percentage of new employees who leave in first 12 months because of unsatisfactory performance	10	8	5	12
4. Wages compared with average	1:1	0.8:1	1.1:1	1.1:1

Figure 9.2   Data matrix for new hiring process benchmark.

Qualitative data also should be collected and analyzed. Some effective ways to present and analyze qualitative data are:

- Word charts (Figure 9.3)
- Work word flowcharts (Figure 9.4)
- Flowcharting (Figure 9.5)

Do not be misled by the measurement data. Just because a location or an organization has better overall performance does not mean that all the activities within its process are world-class. Every process has its strong and weak points. Use all the data you have collected to search out the very best at each activity within the process being studied. Frequently, the world's best activity may not be part of the process used by the world's best organization.

Often, you will find that no one organization has all the right answers, and it will be necessary to combine activities from the organizations studied to establish a new best process. Combining best process activities

	Organization			
	A	B	C	Ours
Forms	One form for new hire and budget change.	Four forms for budget, internal hire, external hire, and offer.	Use computers to process all data. Different screens for budget and hiring.	Two forms: one for budget; one for hiring.
Budget change approval	2d level 3d level Controller Accounting dept. mgr.	2d level 3d level Controller Accounting dept. mgr. Division v.p.	2d level Accounting dept. mgr.	2d level 3d level Accounting dept. mgr.
New employee hiring approval	2d level 3d level Plant mgr. Division personnel Ind. eng.	2d level 3d level Plant mgr. Corporate personnel Personnel manager	2d level Personnel manager	2d level 3d level Ind. eng. Personnel manager

Figure 9.3 Work chart for hiring process.

Process Used by a Manager to Start the Hiring Process

1. Check to see what the overtime has been for the past 3 months. Can get approval for a new hire if overtime in the department has been greater than 60 hours a week for the past 3 months.
2. Prepare job description for new employee.
3. Have salary administration evaluate the new job and classify it.
4. Prepare a payback analysis.
5. Fill out a personnel requisition form and obtain next two-level sign-offs.
6. If added budget is required, fill out a budget variation request and get next-level approval.
7. Prepare a letter of justification and send it to the controller for approval.
8. Send approved personnel request, budget variation request, and job description to personnel placement.

Figure 9.4 Qualitative work word flowchart.

allows you to develop a new standard of excellence and to become the benchmark organization for your process.

It is a good idea to have your measurement chart as complete as possible before you visit an organization. Having this data in hand pinpoints activities in the process with the greatest potential for a major breakthrough at your organization. The primary purpose of these organization visits is to observe activities firsthand and to collect detailed qualitative data and information about planned-for or projected improvements to the process. The data collected not only should provide you with a picture of the current gap existing between you and the best-of-breed, but should furnish you with some insights into future performance. It is essential to use this information to project where the world-class standard will be in the future. It takes time to bring about process change, and even the best process must continue to improve. That is how it got to be the best.

The most dramatic and effective way to illustrate the difference between your process and that of the benchmark organization is by a performance projection chart (Figure 9.6). In this chart, the "now" points are fixed by collected data. You can obtain your organization's past performance data by consulting its history file. Most organizations have registered some performance improvement over the years as a result of normal goal setting. Improvement in productivity of 2 to 5 percent per year is commonplace. Using this established trend, project how the organization will be performing over the next 5 years. Then, based on the benchmarking study, determine the future trends of the benchmark organization using historical data and future projections to define the slope of this line. If no data are available from the benchmark organi-

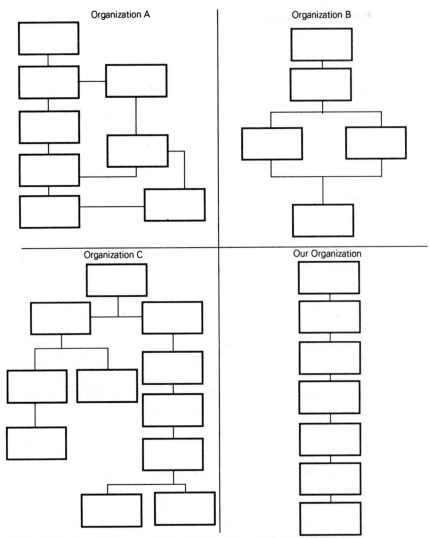

Figure 9.5 Comparable process flowcharts of four different organizations.

zation, adopt the same slope used in projecting your organization's performance. Substituting your organization's slope for the benchmark organization's frequently provides a very optimistic projection. Normally,
the best-of-breed organizations got there because their improvement
rate was much better than that of other organizations.

Analyzing each key measurement will reveal that:

- The gap is in favor of your organization.

Percent of Employees Released
Within First Year of Hiring

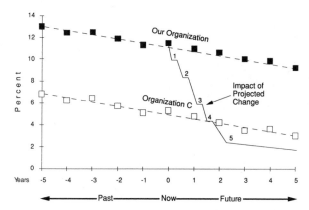

Figure 9.6 Performance projection chart.

- The gap is not in favor of your organization.

  The gap is staying the same.

  The gap is widening.

  The gap is narrowing.

- There is no gap between your organization and the benchmark organization.

When the gap is unfavorable or there is no gap, the measurement is a candidate for an improvement action plan.

Setting Targets and Developing Action Plans

When you start setting targets and developing action plans, qualitative data become invaluable since they tell you what to do to bring about positive change. Most organizations try to implement all the beneficial activities and process changes they have discovered in one massive effort to make a major step-function improvement in their process. This frequently has disastrous effects on the process. Some changes do not work and bring the process to a standstill. Others have no impact, either positive or negative, but add cost to the system.

At this stage, you have a lot of information in your hands. Proceed with caution. Prioritize the potential changes. Pilot them. Measure the results. If the results are positive, pilot the next change. If the results show negative impact or no improvement, remove the change and proceed with the next change (see Figure 9.7).

The best way to prioritize your change activities is to prepare a change impact analysis (Figure 9.8). In this analysis, each of the key

Figure 9.7   Change impact on hiring process.

			Measurements			
Change	A	B	C	D	E	Cost to implement
1	+5	N/A	+2	N/A	N/A	$30,000
2	N/A	+6	N/A	−1	N/A	$20,000
3	+1	+3	N/A	+5	+1	$ 500
4	−3	+1	N/A	N/A	+6	$ 1,000
5	N/A	N/A	+6	N/A	N/A	$ 1,000

N/A = not applicable

Figure 9.8   Change impact analysis.

measurements you hope to improve is listed on the horizontal axis, and each of the proposed changes is listed on the vertical axis. The estimated impact in percentage or actual value that the proposed change will have on the measurement is recorded directly below the measurement. (Be sure that the unit of measurement—percentage or actual value—is constant so that the numbers can be summed.)

The change analysis must consider the total impact of the change. Many changes interactively impact more than one measurement. Often, a change may positively impact one measurement, while negatively impacting others. For example, reducing the density of an epoxy to make it easier to apply also might reduce the strength of the bond, causing the product to fail in the customer's application. At the far right side of the projected activity line, record the estimated cost to implement the change.

When this analysis is used as a base, priorities for changes can be set and their impact assessed. Then prepare a performance projection chart to predict the impact of implementing priority changes. Adding the projected performance change for the priority process changes to the performance projection chart will help determine when and if your process will become world-class (see Figure 9.6).

Updating the BMP

The BMP is a continuous discovery activity. As soon as you stop adding data to the database, it becomes out of date. Public domain data regularly should be added to the database, and someone should be assigned to review these new inputs to identify specific breakthroughs. Every 1 to 3 years, the total BMP should be repeated. This is absolutely necessary. In today's high-technology world, key processes are changing very rapidly. A simple technological breakthrough could revitalize an organization's process and bring a dark horse up into first place overnight.

THE 30 STEPS TO BENCHMARKING RESULTS

Until now, we have talked about the BMP in general. Now let's get down to specifics. The two major parts to the benchmarking process include:

1. The internal benchmarking process, consisting of 14 potential activities (see Figure 9.9).
2. The external benchmarking process, consisting of 16 possible activities (numbered 15 through 30; see Figure 9.10).

INTERNAL BMP

Let's examine each of the internal BMP activities, discussing special points of interest that were not covered earlier in this chapter.

Planning Phase

Activity 1: Identify What to Benchmark.   Identify what products, processes, and/or activities should be benchmarked.

Activity 2: Obtain Management Support.   Obtain management support for the benchmark target products, processes, and/or activities. This support must include project approval and approved human and financial resources.

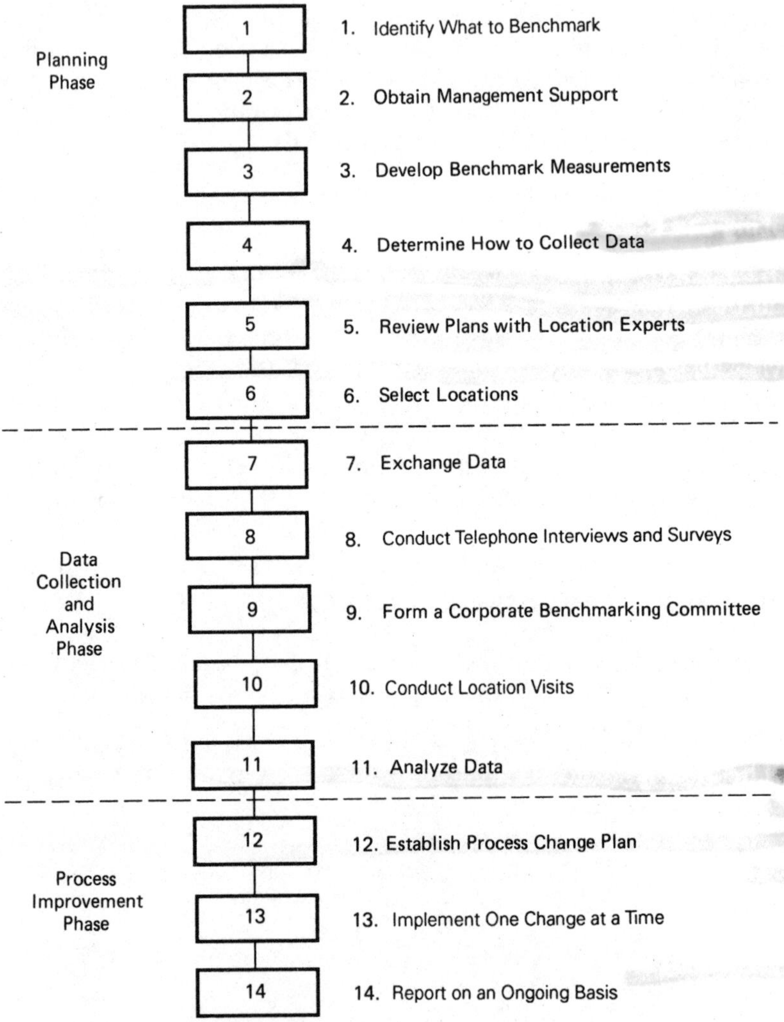

Planning Phase	1	1. Identify What to Benchmark
	2	2. Obtain Management Support
	3	3. Develop Benchmark Measurements
	4	4. Determine How to Collect Data
	5	5. Review Plans with Location Experts
	6	6. Select Locations
Data Collection and Analysis Phase	7	7. Exchange Data
	8	8. Conduct Telephone Interviews and Surveys
	9	9. Form a Corporate Benchmarking Committee
	10	10. Conduct Location Visits
	11	11. Analyze Data
Process Improvement Phase	12	12. Establish Process Change Plan
	13	13. Implement One Change at a Time
	14	14. Report on an Ongoing Basis

Figure 9.9 Internal BMP flow.

Activity 3: Develop Benchmark Measurements. This will involve both qualitative and quantitative data. Develop a measurement matrix using your database.

Activity 4: Determine How to Collect Data. Four of the most commonly used data collection methods are:

- Exchange of process data, procedures, and flowcharts
- Telephone interviews and surveys

- Committees
- Location visits

In most cases, probably all four will be used.

Activity 5: Review Plans with Location Experts.   Search out people who understand the process being evaluated and ask them to:

- Review the data plan
- Recommend other locations that are doing the same or similar activities well
- Identify external organizations to be benchmarked
- Suggest internal and external contacts for information to help identify world-class processes and organizations

Then update the plan to reflect these inputs.

Activity 6: Select Locations.   Select locations to be benchmarked using the updated plan.

Data Collection and Analysis Phase

As mentioned in activity 4, there are four major ways to collect internal information that will be used to start the BMP database.

Activity 7: Exchange Data.   Contact key process people within your location or from other locations and explain what you are trying to accomplish. Ask them to become partners in the BMP. Offer to send them copies of your measurement matrix data, your procedures, and your process flowcharts. Request that they review this data and send back their data in a similar format, along with any comments they have.

Activity 8: Conduct Telephone Interviews and Surveys.   After carefully reviewing the other locations' data, call your contacts and discuss the data you received from them to clarify key activities. Ask if the location would like to be represented on a corporate benchmarking committee for the process being studied. In complex studies, it is best to use a written survey in place of a telephone interview because it allows the individual to acquire exact data, and it documents the response.

Activity 9: Form a Corporate Benchmarking Committee.   The corporate benchmarking committee should review the data collected and discuss in detail the activities identified as being the best, to develop a common understanding. This meeting should provide valuable insights into the areas being studied and highlight areas where additional information is needed.

Activity 10: Conduct Location Visits. The corporate benchmarking committee should hold its meetings at different locations. In conjunction with these meetings, a detailed tour of the process being benchmarked, and other similar processes, should be conducted. Each member of the committee should write process review reports defining strengths and weaknesses observed. Add this information to the benchmarking database.

Activity 11: Analyze Data. The benchmarking committee should construct a process flow diagram including the best processes and practices from all locations. Then the committee should estimate the expected performance if the optimum process is implemented.

Process Improvement Phase

Activity 12: Establish Process Change Plan. Each location should review the total database and develop a performance projection chart that compares the location to the model developed in activity 11. Then it should conduct a change impact analysis to prioritize the implementation activities. In some cases, organizations decide not to make any process changes at this point because the changes may have to be redone when the external benchmark data are added to the database. The decision to implement the change now or later will depend on the projected gains resulting from the change and on the amount of time the change will be delayed while waiting for the external BMP to be completed.

Activity 13: Implement One Change at a Time. Implement the high-priority changes one at a time to evaluate the impact of each change independently.

Activity 14: Report on an Ongoing Basis. Establish a measurement report comparing performance by location. Issue this report every 6 months. IBM, for example, used such a report as early as 1970 to compare the performance of major activities (e.g., quality engineering inspection, manufacturing engineering, personnel, research and development, product engineering, industrial engineering, accounting, and purchasing). Most support activities reported four to ten key ratios.

EXTERNAL BMP

Figure 9.10 illustrates the typical flow of an external benchmarking process. The following material describes the activities in each of these blocks. Additional detail will be provided for areas that were not discussed earlier in this chapter.

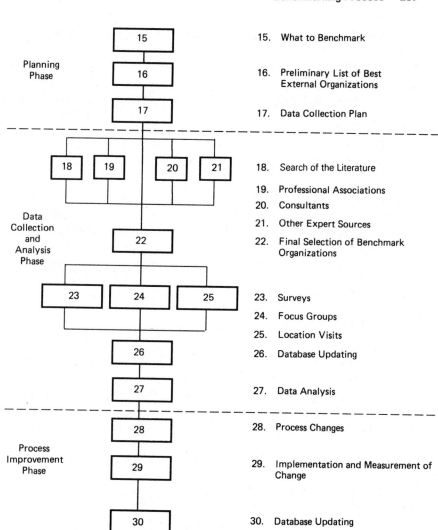

Figure 9.10   External BMP flow.

Planning Phase

Activity 15: What to Benchmark.   Using the benchmarking committee review, and the database, determine whether the total process should be benchmarked or just key activities within the process.

Activity 16: Preliminary List of Best External Organizations.   Using the database, and the expertise of the benchmarking committee, list the or-

ganizations with a world-class reputation. This is a preliminary list that
will be refined later in the BMP.

Activity 17: Data Collection Plan.   There are many ways to collect data
on external firms. The possibilities are limited only by the imagination
and creativity of the members of the benchmarking committee. Some of
the most common are:

- Searches of the literature
- Professional and trade associations
- External experts

A great deal of data about the competition, and about world-class organizations, exist in the public domain. Devote the necessary time and
energy to collecting these data. The more you know about an organization before contacting it, the more apt you are to gain real insights into
its process.

Data Collection and Analysis Phase

Activity 18: Search of the Literature.   A good research librarian can
make this difficult process flow smoothly. Huge databases can be sorted
by key words to identify articles, publications, reports, and books on the
subject under study. Most of these databases list the documents and include a short summary. The organizations maintaining these business
and technical databases can provide you with copies of the desired articles at very reasonable prices. Some of the best sources used in the benchmarking process are:

- Annual reports
- 10K data
- Public magazines
- Industrial journals
- Newspapers
- Periodicals
- Special reports
- Trade publications
- General magazines (e.g., *Newsweek*)
- Association reports and studies
- Association publications
- Books
- Conference proceedings

These data will enhance your understanding of processes used outside
your organization and provide you with key contacts in the best organi-

zations. At a minimum, the search of the literature should cover the last 10 years.

Activity 19: Professional Associations. Do not overlook this important source of information. The number of professional and trade associations is surprisingly high. Your local library has an encyclopedia of associations. You can be almost sure that some professional or trade association has been formed to understand the process you are benchmarking and has experts in the field. A telephone conversation, or a visit to an association's headquarters, is time well spent because such associations may have data and activities that can provide valuable information, such as:

- Conferences
- Association libraries
- Databases
- Field trips
- Reports and publications

Activity 20: Consultants. Consultants continuously search for the best systems, procedures, and practices, and the nature of their work provides them with the opportunity to personally observe the operations of many different organizations in different environments.

A consultant who is actively involved in making an organization's process work (not just teaching) can provide you with an unbiased view of the total process and the individual steps within the process. Frequently, when a process is discussed with a benchmark organization's employees, the employees point out all the good things and let all the bad things go unmentioned. They show you how the process *should* be operating not how it *is* operating. A consultant, on the other hand, can provide you with a view of the process that is not colored by an employee's natural pride in his or her organization.

Another advantage consultants have is that they can act as a third party, providing data without divulging specific sources. Frequently, organizations hesitate to release information that can be compared with that of other organizations because they are afraid it might make them look less effective, efficient, or capable. A consultant provides the veil of anonymity necessary to obtain the full cooperation of the targeted benchmark organization because he or she can ensure that the organization's name will not be associated with unfavorable comparisons.

In practice, holding back the names of the organizations that are not world-class does not present a problem because you are trying to identify the best organizations, not comparing yourself with another organization. The best organization usually is very willing to be listed as such and will give the consultant permission to use its name and document its process. A typical report from your consultant might read, "Ten organizations were benchmarked. Culinar had the best fixture changeover methods and procedures...."

Activity 21: Other Expert Sources. There are a number of other valuable sources of data, including:

- Universities
- Company watchers (brokerage firms)
- Software firms
- Research organizations

Universities and their professors can be an important data source, especially since students who have graduated are a rich source of research information for their former professors. These professors often maintain close contacts with many of their students who have graduated and moved into key positions in business and government. In addition, summer study programs and research projects represent a mine of information available at universities.

Among the most prized sources of information are the brokerage firms that assign employees to collect and analyze data related to an industry and/or organization as their sole task. These people can offer insights into an organization and target key contacts.

Activity 22: Final Selection of Benchmark Organizations. Now is the time to review all the data collected and update your quantitative and qualitative matrix. Identify any voids existing in the collected data. After a detailed analysis of this data, pinpoint the organizations to target for benchmarking and identify key contacts there.

By now you should have reduced your potential benchmarking partners to three to five organizations. Organizations that may have looked good at first will have been dropped from the list because of:

- Unwillingness to share data
- Lack of data
- Existence of better candidates
- Reputation as not the best performer
- Process not comparable to yours
- Communication problems
- Travel costs
- Lack of interest in other organizations

Activity 23: Surveys. You still are not ready to make your first visit to your first benchmark organization. There remains one more approach that should be considered because it uncovers far better data at much less cost. This is the benchmark survey. Surveys administered through a third party will ensure anonymity. They also are an effective and economical way of obtaining qualitative and quantitative data. The advantages of third-party surveys include:

- All parties can remain anonymous.
- The questionnaire is designed to provide all the desired information.
- The person being surveyed has time to collect the required data, making the responses significantly more accurate.
- The questionnaire can be designed to avoid ambiguous answers.
- More extensive data can be collected than are available during a location visit.
- Written data are usually more accurate than oral remarks.

Hard, cold facts are always much more valuable than a best guess, and a best guess is what you often get when you ask a question a person is not prepared to answer.

Developing a well-prepared questionnaire is a science, not a matter of luck. Have your professional people draft the questions to be asked and engage a research professional to design the actual questionnaire.

Model the questionnaire to ensure that you obtain the data you need. The way questions are worded can greatly impact the way they are answered. For example, consider how the same person answered the following questions:

Question 1:   How satisfied are you with software you are using?
Answer: Very satisfied.
Question 2:   Rate the software you are using, compared with the very best, on a scale of 1 to 10, with 1 as bad, 10 as exceptional.
Answer: 5, about average.

Five types of questions can be used in a questionnaire:

1. Multiple choice
2. Scale (as in question 2, above)
3. Written comments
4. Rating
5. Forced choice (e.g., true/false, yes/no)

Do not ask for proprietary information or for data that your organization would not willingly share with benchmark partners.

Surveys can be conducted in a variety of ways:

- Mail surveys
- Telephone surveys

- Face-to-face interviews
- Combination of the above

For best results:

- The individual to be surveyed is contacted in advance to explain to him or her why the survey is being conducted and how the data will be used. He or she should be invited to participate as a benchmark partner. Remember, it takes time and money to fill out a questionnaire, so there should be something in it for the benchmark organization and for the individual doing the work. Frequently, providing a copy of the final report serves as the necessary incentive.
- Mail the survey to the individual. Enclose a self-addressed, stamped return envelope.
- When the survey is returned, review it in detail, and contact the person surveyed to acknowledge its receipt and to clarify any unclear points.
- Add the data collected from the survey to the database. At the end of the survey phase, all the nonproprietary blanks in your qualitative and quantitative matrices should be filled in.
- Forward a copy of the final report, a specially prepared comparison of the specific organization to the other benchmark organizations, and a thank-you letter to the individual.

Activity 24: Focus Groups. Another effective way to collect and evaluate data is to invite interested parties from different organizations and environments to meet and discuss the process under investigation. Often this is called a *focus group* because it focuses its discussions directly upon the process. It is best to have a third party organize and facilitate this exchange of information.

This type of activity not only sets up the standards for the present but also is a very effective way to discuss and evaluate future process changes. Even after the initial study is concluded, it is a good idea to hold periodic focus group meetings. These meetings generate a continuous flow of new ideas and allow the parties involved to benefit from the experiments conducted at the individual locations.

To make the focus group successful, use a third-party facilitator well schooled in group dynamics. Carefully select the participants to ensure that they are at the same technical level and hold the group meetings at a neutral site. Distribute agendas well in advance of the meetings and document the proceedings.

Activity 25: Location Visits. Seeing is believing, and a location visit can be one of the most exciting parts of the total BMP. It is your chance to sit down face to face with your counterparts in other organizations to discuss the processes with which you both are involved. Your team will have the opportunity to see firsthand how the process works in another organization. This tour of your benchmark partner's process provides your team with an opportunity to observe the methods, processes, procedures, equipment, and results on the spot. Because it may be your one chance to get an inside view of what your process could evolve to, you must make the most of every moment you spend in the benchmark organization. A lot of planning should take place before you arrive at the organization's reception desk.

The location visit can be divided as follows:

Planning: By now your database should contain a complete file of the benchmark organization and the process being studied. Each of the people who will be visiting the organization should prepare and study a complete data file for the benchmark partner. You should prepare a questionnaire and a visit agenda. Assign each member of the visiting team a specific process activity to investigate. This provides the best process coverage in the shortest period of time.

Arranging the visit: Do not make the initial contact without careful preparation. The first problem is finding the right contact, preferably by being referred by a mutual acquaintance. Sometimes a business relationship already exists. Another effective approach is to have a professional make contact with his or her counterpart in the benchmark partner firm.

After the contact individual is identified, a phone call is the best way to introduce yourself, and explain the purpose of the project, and how the data will be used. Ask for permission to visit, review the process, and discuss how it functions. Ask if he or she would fill out a questionnaire to provide advance information to help the visit run more effectively.

If the potential benchmark partner agrees, send a letter documenting the discussion and requesting a date for the visit. The letter should include the questionnaire, information about your process, the names and titles of the people who will be making the visit, and a proposed agenda. Never ask the benchmark organization to disclose information that your organization would be reluctant to disclose. If you want to meet with particular people or organizations, communicate this request in the letter. Frequently, it is advantageous to take pictures of the process under study and to audio-tape the discussions. If you desire to do so, be sure to get prior permission.

The visit: For best results, limit the visit team to two to eight people and identify the role of each individual in advance. Also, prepare a list of questions that includes only need-to-know type information.

Review the process flow and the data in a conference room and then tour the process to see firsthand what is going on. Talk to employees. Learn how they feel. Look for activities that set this organization apart.

You may want to divide the visit team into small groups to focus on specific details and individual activities.

Use the time after the tour to share your observations, and review the data collected. Discuss your benchmark partner's plans to improve its present process. This is a key conversation and should help to project future process changes. Before you leave the benchmark organization, extend an invitation to its team to visit your location and observe your process firsthand.

Debriefing: Hold a meeting the same day as the visit to consolidate thinking and document observations. Do not put it off even 1 day. After returning home, each team member should prepare an individual report and add pertinent data to the database.

Following up: Within a week of your visit send a letter thanking your contact and extending an invitation for his or her team to visit your location. Point out the very best practices you observed during the visit. Consider inviting the benchmark organization to join a network that shares best practices on the process being studied on an ongoing basis. Provide the benchmark partner with a target date by which he or she will receive the final benchmarking report.

Final report to benchmark partners: Prepare and distribute a report comparing each benchmark partner to the total population.

Activity 26: Database Updating.   Activities 18 through 25 generate vast amounts of data that must be captured and analyzed. The best way to do this is to constantly update the benchmark database as each activity is performed. Remember the old saying, "The job is not done until the paperwork is complete."

Activity 27: Data Analysis.   This activity was discussed extensively earlier in this chapter under the heading "Data Analysis."

Process Improvement Phase

Activity 28: Process Changes.   Be sure to follow the suggestions given earlier in this chapter under the heading "Setting Targets and Developing Action Plans."

Activity 29: Implementation and Measurement of Change.   Implement one change at a time so that the impact of each change can be evaluated separately. Develop a system that measures the impact of each change on the total process. Many changes can have both positive and negative effects. Any measurement system should evaluate a change's total impact on effectiveness and efficiency as it relates to the total process.

Activity 30: Database Updating.   Your organization has made a major investment in developing an extensive database on the process being studied. Since it is much easier to keep the database current than it is to go back at a later date and try to update it, someone should be assigned to research and update the database on a continuous basis. This individual should include the information available in the public domain, internal evaluations, benchmarking partner results, and focus groups.

SUMMARY

Quality has acquired a new dimension and importance. Today, customers evaluate organizations based on many more factors than they did in the past. Every time your organization's name, products, services, or people come in contact with customers or potential customers, your organization is evaluated and scored.

It will not be long before all organizations provide high-quality products; product quality will be a given. What will then set organizations apart will be their service, their people, and the demonstrated pride these people have in their organization.

Today, it is the sum of all customer contacts that spells quality, and quality is what your customers are buying. Currently, 8 out of 10 customers rate quality as being equal to, or more important than, price in their purchase decisions. Just 8 years ago, only 40 percent felt that this was the case.

Benchmarking is not a new process. It has been used for years to study competitive products. The refinement of techniques in recent years has led to its extensive use in defining the best business processes.

Benchmarking is particularly valuable because it:

- Enables an organization to set challenging, yet realistic, targets
- Provides a process for improvement
- Facilitates prediction of future trends
- Will help your organization improve on the very best so that your organization will itself become a benchmark organization
- Provides information on how to improve

Systematic benchmarking can make your business processes the very best. Without it, you will never truly know how good you are, how good you should be, or how to become the best you can be. Isn't it time you really understood the full potential of your processes?

10

The Beginning

CONTINUOUS IMPROVEMENT PROCESS

As your process becomes the benchmark process for the rest of the world, you have reached a plateau...like graduating from college. But graduating from college is not the end of the learning process. It is just the beginning. Our brain is like the rest of our body; it needs frequent exercise to keep it healthy. Thomas J. Watson, Sr., first president of IBM, stated, "There is no saturation point to knowledge." The same is true of process improvement.

No matter how good you are, how well regarded your products and/ or services are, you cannot stop improving. You cannot stand still. When you do, you really aren't standing still, you are slipping backward because your competition is constantly improving. The very best have to run to stay the very best, because if you are not improving, there is only one direction you can go, and that's down.

Even when your process becomes the very best, you are not at the end of the business process improvement (BPI) cycle. You are at the beginning of the ongoing, continuous improvement phase. You must continue to improve your processes because:

- New methods, programs, and equipment are coming out every day.
- The business environment continues to change, making efficient processes obsolete overnight.
- Consumer and customer expectations change almost daily, making what was outstanding yesterday just meet requirements today and inadequate tomorrow.
- The people within the process develop increased capabilities, providing increased opportunities for process refinement.
- Uncared-for, unattended processes degrade over time.
- No matter how good the process is today, there's always a better way. All you need to do is find it.

When everything is going perfectly, I like to take something out of the process to challenge the team to improve, to give the team another chance to be more creative. The team should ask itself what it would do if customer expectations jumped by 100 percent. How would *you* meet this new challenge? If your head count was cut by 25 percent, what would you do to ensure that customer expectations still were met? Challenge your team and the process to be better today than they were yesterday and better tomorrow than they are today.

We have kept this chapter short because the tools for continuous improvement are embedded throughout the book. This chapter concentrates on the concept of continuous improvement, of never being satisfied with the way things are—an essential ingredient in our quest for excellence, value, competitiveness, and profitability. As managers and employees, we can profit only to the extent that our organization profits. One of today's best opportunities to decrease waste and increase customer satisfaction and our quality of work life is by continuously improving our business processes. What we need to do is continuously go around the quality ring (discussed in Chapter 7, Figure 7.10).

Once you have met customer expectations, you need to set challenge targets for the process. Develop a plan to meet these targets. Implement the plan, adjusting it as more data are available. When the challenge targets are reached, you will need to take time out to rejoice and reward the team. The content of the reward should be commensurate with the magnitude of the accomplishment. In some cases, it may be coffee at a team meeting or a cake to recognize a milestone. In other cases, it may result in a financial reward when the accomplishments have a major impact on the business. As soon as the team has been recognized, the cycle starts over again. New targets are set; new plans are developed. These plans are implemented. Targets are met. The team is rewarded. Around and around the quality ring you go...always getting better, always improving. For the organization to keep its competitive position and for us to keep our jobs, we must always get better and better. It's the law of survival.

But teams can't do it by themselves. We need to get everyone actively involved. From the boardroom to the boiler room, from the salesperson to security, everyone must be searching for ways to improve the process.

CONSTRUCTIVE DISSATISFACTION

We often ask clients if they would like to have dissatisfied employees working for them. Most of the time they say No. We want dissatisfied employees working for us, but only if they are what can be called *constructively dissatisfied*. Constructively dissatisfied employees are ones who believe that things can be better, and they are the ones who can bring about the required change. They are willing to challenge the systems, even make management justify why it is doing things the same old way. At IBM, these types of people were called *wild ducks*. Management spends too much time clipping the wings of the wild ducks, when it should be encouraging them to fly. They are much harder to manage than the "yes people," but the wild ducks are our future. They are the ones who originate the breakthroughs that make organizations world-class.

HARRINGTON'S WHEEL OF FORTUNE

The BPI concepts in themselves will help any organization improve, but for the best, long-lasting results, these concepts must be supported by a complete improvement process that involves all facets of the organization (Figure 10.1).

As you look at the wheel of fortune, you will note that the outer ring that holds the wheel together consists of management leading a process of unending change directed at continuous improvement. The wheel focuses this change process on making the organization more effective and efficient (the hub of the wheel). The spokes of the wheel make up the principles that are required to bring about continuous improvement.

■ *Customer focus.* Customer focus is at the top of the wheel because it is the most important principle. Understanding customer needs and expectations is essential to staying in business. Your organization needs to be so close to customers that you can realize their present and future needs even before they recognize them.

Figure 10.1 Harrington's wheel of fortune.

- *Planning.* Excellence does not happen by accident. It requires a well-thought-out, well-communicated plan that is embraced by everyone. This plan must be based on a shared vision of how the organization will function and how the quality of work life will improve, while providing output to the external customer that surpasses his or her expectations.

- *Trust.* Management must trust the employees before it can earn the trust of the employees. To excel, both parties have to trust each other. Employees will never willingly identify and eliminate waste until they trust management enough to be sure their own jobs will not be eliminated as a result of their suggested improvements.

- *Standardized processes.* Real improvement occurs when everyone is performing an activity in the same way so that the results are predictable. When different people approach the same task in many different ways, the results are difficult to control or improve. Standardization is a key step in the improvement cycle.

- *Process focus.* As we strive to continuously improve, we need to focus our efforts on improving the process, not on who caused the problem. The improvement process must define what went wrong with the process that allowed the problem to occur so that the process can be changed to prevent the problem from ever recurring.

- *Total participation.* No one in the organization is immune from the continuous improvement process. All people have to have a responsibility to improve the way they are working and to help the team they are part of be better. The time when employees were hired for their physical abilities alone is gone. We need the ideas and cooperation of all the team members if we are to excel. Everyone must be involved and be actively encouraged to participate in the improvement process.

■ *Training.* Would you consider not putting oil in your car or not maintaining the equipment in your manufacturing process? The answer has to be No. Your most valuable asset is your people. Training is the maintenance of your human resources. Training is an investment in the future of your organization. Training is not costly...it is ignorance that is costly. Organizations are undergoing rapid change in the way they operate and the way people think, talk, and act. This change process must be supported by an aggressive training program that reinforces these changes and provides growth opportunities for your employees.

■ *"Us" relationships.* All members of the organization need to realize that they are part of the total organization. All individuals have customers to whom they provide output and suppliers who provide input to them. Our success, our growth, and our rewards are based on how well the total organization performs. Everyone has to stop thinking about *my* job, *those* people, and *management versus employees*. It can no longer be a *we* and *them* type of operation. The organization is *us*. Working together, we can make it better, make it grow, and make it prosper; and as the organization prospers, all of *us* will prosper.

■ *Statistical thinking.* We can no longer run our complex business by our best guess. We need hard data. We need to know how much faith we can put in the data we receive. We need to know our options. We need to know the probability of success. Our businesses are too complex, and the difference between success and failure too small, to continue to fly by the "seat of our pants." Sure, there will always be some judgment involved in many final decisions, but we should be able to quantify the risks involved in these judgment decisions.

■ *Rewards.* Rewards and recognition are an essential part of the improvement process. They reinforce desired behavior and visually demonstrate management appreciation of a job well done. To accomplish the desired result, a comprehensive reward and recognition system needs to be developed, one that allows management the latitude to be creative. Everyone hears *thank-you* in a different way. In addition, the reward must be based on the magnitude of the contribution. You need both team and individual rewards. At times, a simple thank-you is appropriate; on other occasions, financial reward is more appropriate.

Take, for example, an employee who has worked hard for the last 3 months and has come up with an idea that saved the organization 1 million dollars. Her manager, in an effort to reward the employee, walks up to her, shakes her hand, and says, "That was an outstanding job. Keep up the good work, Jane." Jane replies, "Thanks, boss. I will try."

But what Jane is really thinking is, "I saved the organization 1 million dollars, and all I get is a thank-you! That's the last time I am going to knock myself out for this organization." There is a time for a pat on the back, and a time for a pat on the wallet.

PUTTING IT ALL TOGETHER

In going around Harrington's wheel of fortune, we end up at customer focus, and quite appropriately so, because we can never get far away from the customer. As John A. Young, president of Hewlett-Packard, put it, "Satisfying customers is the only reason we are in business." Harold A. Poling, president of Ford Motor Company, stated, "Continuous improvement in everything that is done is what it will take to continue to satisfy the customer."

In Chapter 2, we described the way to start your organization on its improvement journey. It is a long, long journey, with no end, a race with no finish line. Some people never start down this long road because they see no end. Others start jogging down the road and stop under a shady tree, never to reenter the race. Others get up every day, get back on the road, and make real progress. These are the people who make a difference. They make a difference to themselves, their families, their organizations, and their countries. Please join me in my long run for continuous improvement!

Appendix

Interview Guidelines*

INTERVIEWING SKILLS

The ability to gather highly accurate and detailed information is critical to the success of almost every business process improvement (BPI) technique described in this book. These techniques depend on a knowledge of the internal and external customer requirements, the current process activities and flow, and supplier capabilities. The most commonly used method of gathering needed information is the one-on-one, interactive interview.

Interviews are used to collect information from internal and external process customers, people who execute the process at the task level, process owners and managers, and suppliers to the process. They are also used in benchmarking to acquire information on best industry practices. Customer market research is based on one-on-one interviews

---

*This appendix was written for this book by Daniel M. Stowell, D. M. Stowell & Company, Stamford, Connecticut.

as well as telephone surveys and focus groups, all of which demand effective interviewing skills.

The purpose of this appendix is to provide the quality professional, manager, or process team member with the specific skills needed for successful interviewing. The focus will be on the one-on-one face-to-face interview. However, many of the concepts covered will be relevant to some degree to telephone surveys, focus groups, and even direct-mail questionnaires.

Planning the Interview

Information gathering occurs as an activity or step within a process. The objective of information gathering is to provide the data required to successfully execute that process. Within the context of this book, the objective of the process is generally to improve the quality—the effectiveness and efficiency—of business processes. Processes include functions such as billing, inventory, marketing, payroll, accounting, and planning.

As with any process or activity, the first step in interviewing is to identify the customer, the individual who will use the information to be gathered. The next step is to determine the customer's requirements, including what information is needed, at what level of detail, in what format, and when.

The best approach for gathering that information should be selected. Methods include one-on-one interviews, focus groups, telephone surveys, and direct-mail questionnaires. With the advent of local and remote computing networks, surveys can even be conducted electronically.

The information required and the collection method will be determined to some extent by the specific quality improvement process being used. For example, the process analysis technique (PAT) requires the use of one-on-one interviews, while benchmarking uses almost every approach mentioned above.

Interview Guide Preparation

The next step is to create an interview guide. Key to the success of the interview, the guide serves several functions. It includes an introduction designed to cover all the preliminary information needed by the interviewee. It provides a logical and orderly sequence to the interview and ensures that questions are phrased in an easy-to-understand and consistent manner. It may also include a glossary of industry- or process-specific terms for use during the interview. Most important, it provides a structure to ensure consistency across all the interviews.

The introduction will be read or paraphrased by the interviewer at the beginning of the interview. Some items which may be included in the introduction are:

1. Why the interview is being conducted
2. Who authorized the interview
3. Who else is being interviewed (by title, job type, etc.)
4. How the interviewee was selected and by whom
5. How the information will be used
6. Whether the person will be anonymous
7. Whether the person will be quoted in the summary findings
8. What feedback the person will receive
9. How the person might participate in the outcome of the process
10. What is in it for the interviewee
11. Why highly detailed, accurate information is important to the success of the interview
12. How the person plays a key role in an important process
13. Request for permission to tape-record the interview
14. Invitation for the interviewee to ask questions about the interview process itself

Generally, informational questions are located next in the guide, although some interviewers prefer to put them at the end. These questions include requests for information such as interviewee name, job title, time in the job, time in the business, and department name and number.

The third section consists of the questions relating to the subject of the interview. In constructing these questions, keep in mind the educational level and cultural background of the interviewees and use words and sentences at a level that they can easily understand and with which they are comfortable. If a series of interviews covers many levels in an organization, consider the use of multiple guides, each tailored to fit a specific level.

Avoid the use of jargon if possible. If industry-specific or process-specific terms are necessary, be sure that the interviewee understands the accepted definitions. Include the definitions in an appendix to the guide for reference during the interview.

When the guide is complete, it may be appropriate to conduct a few pilot interviews. Two to four such interviews should be sufficient to determine how well the guide flows and how long the interviews will take. Additional topics may be identified, and redundant questions eliminated. After the pilot interviews are completed, the interview guide can be revised and put in final form.

Selecting the People to Interview

Selecting the appropriate people to interview is key to the success of the quality improvement process. Specifically how to go about the selection depends on the improvement process being employed.

For example, if the interviews are in support of PAT, selecting candidates is relatively easy. Find out from the process managers who knows the process best. Consider who has been doing it longest and who is considered by the task-level people to be the most knowledgeable.

If the interviews are for market research purposes, selecting the appropriate interviewees becomes more difficult. Look for the people in the customer organization who make the buying decision, those who influence the buying decision, and those who use the product. All may have information appropriate to the interview.

Locating good interview candidates for benchmarking is very challenging. Robert Camp devotes a large part of his book *Benchmarking* to the process of identifying companies to contact and interview.

Scheduling the Interview

Scheduling interviews well in advance will give the interviewees the opportunity to fit the interviews into their calendars and to prepare adequately. When you are carrying out process analysis, it is best to do the interviews in the sequence in which the process is executed.

Individual interviews can run from 30 minutes to 2 hours or more. Experience will tell you how much time to allow for each interview. Generally, interviewing managers takes more time than interviewing task-level people, and executives take more time yet.

If you are unsure how much time will be required, plan more than you think will be needed. It is better to finish early than to run over the allotted time. The interviewee may have to return to his or her regular job assignment or may have another commitment.

Schedule only the number of interviews within a day that can be comfortably completed. Allow adequate time between interviews to make notes and relax for a few minutes. Interviewing takes a great deal of energy. Planning eight 1-hour interviews in a day will result in an interviewer exhausted by noon and ineffective in the afternoon sessions.

Location of the Interview

Select an interview site away from the subject's place of work if possible. This helps eliminate interruptions and puts the interview on neutral ground.

Hold the interview in a location where the person feels at ease. A third-shift shipping clerk will feel comfortable, and probably not distracted, in a meeting room or office near the work area than he or she would in a walnut-paneled boardroom.

Avoid holding interviews in the interviewer's own office. This is particularly true if a manager is interviewing his or her own employees. Being interviewed in their manager's office tends to put the employees on the defensive and make them more guarded in their responses.

A critical part of the BPI interview is firsthand observation of the activity being performed. As a result, at least part of the BPI interview will often be conducted in the work area.

Inviting the Interviewees

The next step is to inform the interviewees and invite them to participate. This can be done by the person's manager, directly or by letter. It can also be done in an informational meeting which includes the managers and the people to be interviewed.

Ideally, a senior member of the management team will be involved in the invitation process. This may be done by a letter to the interviewees or through participation at the announcement meeting. Involvement by a senior manager demonstrates that top management supports the improvement process and the interviewee's participation in it.

However they are informed, the invitees need to understand the purpose of the interviews, when they will begin and end, where they are to be held, and what preparation is required of them.

What's in It for the Interviewee

The people being interviewed are making a significant contribution to quality improvement, providing valuable information and insight. They need to understand the benefit to them to make this effort.

If the interviewees are people who work within a business process, the opportunity to improve the process and be recognized for it may be enough. If they are customers of the process, the chance to improve their supplier's product or service may be sufficient. If they are suppliers, the opportunity to shape the process and to ensure that they can provide the appropriate input is of value.

In some cases, something more tangible may be called for. If the results of the interviews are not confidential, the participants may get a copy of the final report. This may be in the form of a summary of the complete findings.

In cases of customer market research interviews and focus groups, the people being interviewed often receive payment for their time.

They may also receive gift certificates or merchandise. If the interviews take place around mealtime, a free meal may be provided.

Appearance in the Interview

Dress and personal appearance in an interview are important. Clothing worn to an interview should be similar to that of the subject or slightly more formal. For example, don't interview the chief executive officer in casual attire, or even a sports coat and slacks, unless that is the way she or he dresses.

On the other hand, don't appear on the plant floor during the third shift wearing a dark blue pinstripe three-piece business suit. Your interview subjects will probably be overpowered by the clothing and will not be willing to open up to you.

Tape-Recording the Interview

Tape-recording interviews is very useful. It allows the interviewer to devote full attention to what is being said without having to take extensive notes. In addition, other people can listen to the interview later.

High-quality microcassette recorders are very good. They are very compact and easy to carry and handle during an interview. A single cassette can store a 1-hour interview.

Get a recorder with a black case. Within a few minutes after the interview begins, the interviewee will be unaware of its presence. (This is why professional photographers use cameras with black cases rather than polished metal.)

This may sound very basic, but be sure to learn how to use the tape recorder thoroughly before starting interviews. Almost every professional interviewer has lost valuable information by becoming confused as to which direction to move the pause control or the voice-activated switch.

Take extra tapes and batteries to the interviews. Use long-running tapes to avoid stopping midinterview to change tapes. The 60-minute type needs only one flip of the tape during a 1-hour interview.

The Interview

The first step in a successful interview is to arrive on time and to be prepared to start on time. A late arrival creates a very negative impression and wastes valuable interview time. Promptness is particularly important with internal executives and customers. If at all possible, try to arrange for all phone calls and other interruptions to be held.

When meeting an interviewee, greet him or her warmly and with a smile and then invite the person to sit down, relax, and feel comfortable. The objective is to gain rapport, to get the person's trust and confidence.

When the interviewee is seated and comfortable, begin with a discussion of the interview guide and how you are going to use it. Then read or paraphrase the introduction. Give the subject the opportunity to ask any questions about how the interview process works, how the information will be used, etc. Doing this allows the interviewer to get to know the person better and begin to build stronger rapport.

Be sure to point out that it is important that you fully communicate with each other. Ask the interviewee to interrupt and request clarification, if you use industry terminology or jargon that the subject does not understand.

If the interview is to be taped, this is an appropriate time to ask permission. Explain that the tape recorder will be used so that you can pay complete attention and not need to take detailed notes. Show the interviewee the tape recorder, turn it on, and then place it aside. Thereafter, handle it only to change the tape.

Incidentally, in more than 12 years of interviewing, the author has never been turned down in a request to record an interview. If an interviewee is uncomfortable, or asks that the tape not be used or be turned off during the interview, do so immediately. Put the recorder away and take lots of very good notes.

Now begin questioning, following the sequence called for in the interview guide. Remember, however, that the guide is just that. It does not necessarily have to be followed step-by-step. If the interviewee gets off the track, but the information is useful, go along with the discussion. The guide is there as an aid to ensure that all the required points have been covered in the interview.

Keep in mind that the objective is to gather information as the interviewee perceives it, not as it really is or as the interviewer perceives it. There is always a temptation to correct an interviewee who responds with information the interviewer knows to be incorrect. The interviewer must resist the temptation to correct the person and must remember to keep gathering information.

If the interviewee has incorrect information, the interviewer should wait until the interview is over and then turn off the recorder and provide the correct information. At this point, you can recognize the person's perception and then ask a question like:

> I understand that you don't believe the firm provides good service. What are some examples of good service? What if you found out the firm did provide that service? How would you like to have found that out that the service was available?

If the subject says something that is not entirely clear, stop and ask for clarification. It may be an important point, or something on which future questions may be based.

Give your full attention to the subject. Deal with the subject as an equal. This will require flexibility since you may be interviewing first-line production and administrative employees as well as executives.

In addition to starting on schedule, make every effort to end the interview within the scheduled time. If the meeting cannot be completed on time, ask permission to continue or schedule another meeting to complete the interview later.

When the interview is complete, summarize the key points and give the interviewee an opportunity to make any closing comments. Be sure to tell the interviewee what happens next. Thank the interviewee for his or her time and the information which has been provided.

Interview Follow-Up

At this point you have a tape and possibly some written documentation and flowcharts. Follow up with a thank-you letter to the interviewee and, if appropriate, to the person's manager or to the person who set up the interview.

This last point is key. Like any other skill, interviewing takes practice. Do it over and over again. Watch to see the results you get as you phrase questions in different ways. Be sensitive to the reaction to everything you do. Continue to improve your skills by observing results, changing your technique, and doing it again.

GAINING RAPPORT DURING THE INTERVIEW

High-quality communication begins by establishing rapport between the parties involved. Most sales texts suggest getting rapport with a potential customer as the first step in an effective sales call. Establishing rapport is an equally important skill in the interview process.

Rapport is defined by the *New Webster's Dictionary* as "harmony" or "affinity." People who have rapport with one another feel comfortable together and tend to be open and honest in their communications. This is certainly a desirable situation in the interview process.

The question then becomes, "How can the interviewer gain a state of profound rapport with the interviewee in a very short period of time?" The traditional approach is to try to get to know someone, to find a common ground. It entails asking questions such as Where are you from? What do you do for a living? What are your hobbies? and even What is your astrological sign? Although, given enough time and some common ground, this will sometimes work, it is not the fastest and most elegant approach.

Think of a time when you saw someone, or when you just exchanged a few words with someone and immediately felt comfortable with that person. You might even have felt as if you had known the person for a long time. Did you ever wonder what it was that created this feeling of rapport so quickly?

In the early 1970s, John Grinder and Richard Bandler developed a communications model called *neurolinguistic programming (NLP)*. They did this by studying several prominent therapists and communicators. Some NLP concepts explain how a high level of rapport can be established very quickly. Grinder and Bandler also described NLP in such a way that it can be easily learned and applied.

The first concept is called *matching and mirroring*. The idea is to help the interviewee to feel comfortable by matching his or her posture and breathing patterns. We have already discussed the idea of dressing at approximately the same level as the subject. This is a form of matching.

Observe the interviewee's posture. Is it erect, relaxed, or slouched? Is the person sitting with arms crossed or open? At what rate is he breathing? Does she use her arms and hands expressively? It is not necessary to mirror the person exactly. Just pick some specific characteristics and match them.

For practice, watch people at the next meeting or party you attend. People match and mirror unconsciously. They are not even aware of what they are doing. Then practice it yourself.

Next, let's look at speech patterns. People represent their outside world by creating a map or model inside their brain. This model of the world is represented by pictures, sounds and words, and feelings. However, people generally favor one of the three methods of representation and are less aware of the other two. They have a favored *representational system* which is visual, auditory, or kinesthetic.

There are several ways to determine a person's favorite representational system. The easiest is to listen to the predicates they use most frequently. Visual people will say things like "I see your point" or "I've got a big picture of that." Auditory people will say "I hear you," "That really clicks with me," or "It sounds good to me." Kinesthetic people will use feeling-related words and will say things like "I am comfortable with that" or "I really need to get a handle on that."

Another indicator of a person's favored representational system is their rate of speech. Visuals speak very rapidly since they are describing pictures in their head. Auditories speak in a more measured pace. Kinesthetic people tend to speak very slowly.

People are most comfortable with people who are like them. Therefore, if you are interviewing someone who is visual, you will be better received if you speak quickly and use visual predicates. If your subject is auditory, use auditory predicates and speak clearly, distinctly, and at a medium pace; and if the interviewee is kinesthetic, speak slowly and use predicates which relate to feelings.

Using physical and language matching will help build rapport with the interviewee very quickly and help get the best information.

INTERVIEW TECHNIQUES FOR ACHIEVING HIGHEST-QUALITY INFORMATION

In addition to the rapport-building skills already discussed, NLP also describes questioning techniques which can be used to elicit the highly detailed and accurate information needed for quality improvement. These techniques are based on a model that Bandler and Grinder call the *Meta model.*

To understand these skills, it is important to understand that the language we use is only a model of what is represented in the human brain. The representation in the human brain is made up of pictures, sounds, and feelings which represent a person's experience of the real world.

As questions are asked during an interview, these pictures, sounds, and feelings are described by the interviewee in words. Of course, the words alone do not provide a complete representation. It is up to the interviewer to know whether the words represent enough of the interviewer's model to be sufficient for the purposes of the interview.

If the words are not sufficient, the interviewer must know how to ask additional questions to create a richer description. This description will be about the needed customer requirements, internal business process, supplier's ability to provide the required input, executive's need for management data, or employee's need for feedback.

The following examples demonstrate how the Meta modeling technique can be used to gather additional information. Of course, in many of the examples other questions could be asked as well. Also, for ease of comprehension, the phrasing of the questions in these examples is simple and direct. The questions may be softened in a real interview.

Omissions

Often interviewees omit information from their responses. For the interviewer to make sense of a sentence, he or she either must supply the omitted information inside his or her brain or must recognize that something is missing and ask for more information.

Here are some examples of omissions and the interview question designed to elicit the omitted material. (In the following examples, Q's are the interviewer's questions and R's are the interviewee's responses.)

Q. What is the biggest problem regarding the billing process?
R. The process costs too much.
Q. Too much compared to what?
Q. What prevents us from responding to the customer's request for additional information on the invoice?
R. The information is too expensive to provide.
Q. Too expensive compared to what?
Q. What prevents the customer invoices from getting out on time?
R. The computer system was down too often.
Q. How often was the system down? What is an acceptable level of downtime?

Notice that in each example, useful information was missing from the interviewee's response and an additional question was asked to get that information. In a real situation, it might be necessary to ask several questions to get the level of detail required.

Unspecified Nouns

Interviewees frequently use nouns that are not specific enough. The word *vehicle* is less specific than *car, car* less than *Ford, Ford* less than *Taurus*. Less specific yet are words like *this, it,* and *they*. Here are examples of typical interview responses and the questions designed to get the additional information required.

Q. What is the biggest problem in the department?
R. The monthly report is out late again.
Q. Specifically, which monthly report is out late?
Q. What prevents the department from making its budget?
R. The workers are just not productive.
Q. Specifically, which workers are not productive?
Q. What is the next step in this process?
R. They bring the invoices to the data entry clerks.
Q. Who, specifically, brings the invoices?

Unspecific Verbs

Interviewees often use relatively unspecific verbs, although some may be more specific than others. *Move* is less specific, for example, than *carry*. To get more detail on unspecific verbs, simply ask "How?"

Q. What is the first thing that happens in the billing process?
R. The shipping notices come to the billing department.
Q. Specifically, how do the shipping notices come to the department?

Q. How are delivery schedules established?
R. Manufacturing forces us to conform to their build schedule.
Q. Specifically, how do they force you to conform?
Q. How do you know when a product ships?
R. Shipping and receiving lets us know.
Q. Specifically, how do they let you know?

Words Implying Necessity

Interviewees often use words that imply judgments, such as *should, should not, cannot,* and *must.* These imply a lack of choice. It is often useful to get additional information to understand what is behind these words and phrases. This understanding will help expand the number of choices available. Here are some examples of interventions that can be used to expand the interviewee's thinking and possibly generate additional options for process improvement.

Q. Can your department do the commission rate check function as part of the revised sales payroll process?
R. No, we can't do that here.
Q. What prevents you from doing it?
Q. The customer needs to know the shipping date within 1 week of order submission. Can you provide it?
R. We should not give them that information.
Q. What would happen if you did? What keeps you from doing it?
Q. Checking part numbers against the master file is now done by computer. Can you stop doing it in this department?
R. We really should continue to do it.
Q. What would happen if you stopped?

Universal Quantifiers

Words such as *never, always, all,* and *everybody* are universal quantifiers and are usually broad generalizations. There are exceptions in almost every process or customer requirement. When an interviewer hears one of these universals, a flag should be raised. Here are suggested interventions to get the additional information required.

Q. Do you check the salesperson's employee number on each order to ensure correct commission payment?
R. No. The employee number is always on the order form.
Q. Always? Without checking, how are you certain?
Q. Mr. Customer, what is your requirement for color coding on the packaging?
R. We will never have a need for that.

Q. Can you think of a time in the past when it might have been convenient to have color coding?

Q. How do the employees feel about implementing the new process?

R. They are all enthusiastically in favor of it.

Q. *All?* Who might possibly have their job affected in such a way as to not favor it?

Rules

Rules are generalizations; they are made by the interviewee about what is right or wrong for everybody and are frequently judgmental. In fact, they are usually what interviewees believe to be true for themselves and true, in general, for everyone. The intervention is to ask "For whom?" Here are examples.

Q. How do you feel about the proposed new process?

R. We do it the right way now.

Q. Right for whom?

Q. How does having marketing quote delivery dates affect scheduling?

R. It is wrong for them to quote the dates.

Q. Wrong for whom?

Q. What is your reaction to the new employee recognition system?

R. It is wrong to recognize employees for doing their jobs.

Q. Wrong for whom?

Nominalizations

Words that have been transformed from verbs into nouns are called *nominalizations.* Changing a verb to a noun causes something that is ongoing to become something that is fixed. For example, the word *relationship* is a nominalization. The verb is *to relate.* It is much easier to think of changing the way we relate to a person, another department, or a union than it is to change a thing called a *relationship.*

Identifying a nominalization is easy to do. The first test is, "Can you put it in a box?" You can easily put a hat, a pencil, a computer, or even a car in a box if the box is big enough. It is difficult to put a relationship, a process, a projection, or a failure in a box.

The second test is to put the word in the phrase *an ongoing,* and if it fits, it is a nominalization. For example, *an ongoing house* does not fit, but *an ongoing process* does. Here are some examples of how nominalizations come up in normal business discussions and how to reframe them as verbs.

Q. What was the result of your Quality Circle work?
R. I didn't get any recognition.
Q. How would you like to be recognized?
Q. What prevents you from implementing the new inventory system?
R. We have a very poor relationship with the union.
Q. How can you improve the way you relate to the union?
Q. How are the employees reacting to the new system?
R. There is a great deal of confusion as to their part in the implementation program.
Q. How can the announcement material be changed so that they will not be confused?

Cause and Effect

Interview subjects sometimes infer that something done by someone else, another group of people, another department, or customers causes them to act in a certain way or to experience some inner feeling. The question to clarify the situation is simply to ask, "How does X cause Y?" Here are some examples:

Q. What is the primary reason for the excessive error rates?
R. The union workers cause most of the quality problems here.
Q. Specifically, how do the union workers cause the problems?
Q. What causes the high absentee rate in the department?
R. The management here makes us sick.
Q. How does management make you sick?
Q. The employee opinion survey indicates very low morale in this department. What seems to be the problem?
R. The salespeople keep everyone upset here.
Q. How do they keep things in a turmoil?

There is one more NLP skill which is needed to tie together all these techniques. In NLP terminology it is called *sensory acuity*. Sensory acuity refers to a very high level of awareness of what is going on in the interview process.

Pay close attention to the physical posture and breathing patterns of the interviewee in order to match and mirror. Be very sensitive to words and sentences in order to use the Meta model questions. Notice what part of the information is being supplied by the interviewee and what part is being implied by the interviewer.

These are the NLP Meta model techniques that can get the high-quality information needed from interviews with task level employees, managers, customers, and suppliers. They require practice and a high level of awareness.

Many readers may already use these communication techniques, some very effectively. It is hard to imagine an executive, a manager, or a professional being effective without using most of these skills.

Our objective in describing them here is to heighten the reader's awareness and to encourage the reader to practice them formally and to observe the results. Coupled with the rapport-building skills, they will aid the reader in becoming more effective in quality improvement efforts.

AFTER THE INTERVIEW

One of the key activities in the BPI walk-through process is to observe the activity being performed. Immediately after the interview, the interviewer and the interviewee should go to the work area to observe the activity discussed in the interview. Observing the individual tasks being performed will stimulate additional questions. As Dr. H. James Harrington puts it, "You never really understand the activity until you do it yourself. If that isn't possible, the next best alternative is to observe the activity while it is being performed, and ask a lot of questions."

Index